PLASTICS ADDITIVES

PLASTICS ADDITIVES

AN INDUSTRIAL GUIDE

Third Edition

Volume II

by

Ernest W. Flick

NOYES PUBLICATIONS
WILLIAM ANDREW PUBLISHING

Norwich, New York, U.S.A.

Copyright © 2002 by William Andrew Publishing
No part of this book may be reproduced or
utilized in any form or by any means, electronic or mechanical, including photocopying,
recording or by any information storage and
retrieval system, without permission in writing
from the Publisher.
Library of Congress Catalog Card Number: 2001088672
ISBN: 0-8155-1472-7 (V.2)
Printed in the United States

Published in the United States of America by
Noyes Publications / William Andrew Publishing
13 Eaton Avenue
Norwich, New York, 13815
1-800-932-7045
www.williamandrew.com
www.knovel.com

10 9 8 7 6 5 4 3 2 1

Library of Congress Cataloging-in-Publication Data

Flick, Ernest W.
        Plastics additives: an industrial guide / by Ernest W. Flick. --
3rd ed.
            p.   cm.
        Includes index.
        ISBN 0-8155-1472-7
        1. Plastics--Additives. I. Title.
TP1142.F58   2002   668.4'11--dc20
                                            2001088672
                                            CIP

                            ISBN: 0-8155-1464-6 (V.1)
                            ISBN: 0-8155-1472-7 (V.2)

To
Spencer Taylor

Preface

The Third Edition of this useful book is divided into three volumes. Volume II describes almost 1,400 plastics additives which are currently available to industry. It is the result of information received from 74 industrial companies and other organizations. The data represents selections from manufacturers' descriptions made at no cost to, nor influence from, the makers or distributors of these materials. Only the most recent information has been included. It is believed that all of the products listed here are currently available, which will be of utmost interest to readers concerned with product discontinuances.

Plastics additives, a complex and growing group of minerals and chemical derivatives, account for 15 to 20% by weight of the total volume of plastic products marketed. They are produced by a large number of companies for a wide variety of customers and needs. Growth of the use of these additives, which impart one or more desirable properties to resins, is relatively strong, about 3.5 to 4% per year. Environmental constraints, however, have imposed rigorous performance requirements on many products, thus placing added expenses on the development costs of these materials.

Plastics additives are divided into logical sections in three volumes.

Volume I:

    I. Adhesion Promoters

    II. Anti-Fogging Agents

The Table of Contents of each volume is organized to also serve as a subject index. Each raw material is located in the section which is most applicable. The reader seeking a specific raw material should check each section which could possibly apply. In addition to the above, two further sections are included in each volume: a list of Suppliers' Addresses and a Trade Name Index. These will be extremely helpful to the reader.

The following information is supplied for each product, as available, in the manufacturer's own words:

Company name and product category.

Trade names and product numbers.

Product description: a description of the product's main features, as described by the supplier.

The three volumes of this book will be of value to technical and managerial personnel involved in the preparation of products made with these plastics additives as well as to the suppliers of the basic raw materials.

My fullest appreciation is expressed to the companies and organizations who supplied the data included in this book.

March, 2001   Ernest W. Flick

NOTICE

To the best of our knowledge the information in this publication is accurate; however, the Publisher does not assume any responsibility for the accuracy or completeness of, or consequences arising from, such information. This industrial guide does not purport to contain detailed user instructions, and by its range and scope could not possibly do so. Mention of trade names or commercial products does not constitute endorsement or recommendation for use by the Publisher.

In some cases plastics additives could be toxic, and therefore due caution should be exercised. Final determination of the suitability of any information or product for use contemplated by any user, and the manner of that use, is the sole responsibility of the user. We strongly recommend that users seek and adhere to a manufacturer's or supplier's current instructions for handling each material they use.

The Author and Publisher have used their best efforts to include only the most recent data available. The reader is cautioned to consult the supplier in case of questions regarding current availability.

Contents and Subject Index

SECTION X: IMPACT MODIFIERS

SECTION XI: INITIATORS

SECTION XII: LUBRICANTS

SECTION XIII: MICAS

SECTION XIV: PIGMENTS, COLORANTS, AND DYES

SECTION XV: PLASTICIZERS AND ESTERS

SECTION XVI: PROCESSING AIDS

SECTION XVII: RELEASE AGENTS

SECTION XVIII: SILANES, TITANATES AND ZIRCONATES

SECTION XIX: SLIP AND ANTI-BLOCKING AGENTS

Section X
Impact Modifiers

Atofina Chemicals Inc.: DURASTRENGTH High Efficiency Impact Modifiers:

Durastrength 200:
    Main Features: excellent impact
                   excellent processing
                   good weatherability
    Applications: siding substrate
                  siding capstock (medium to high gloss)
                  profile (high gloss)
                  specialty pipe
                  injection molding
                  foam profiles
                  flexible applications
    Chemical Description: modified acrylic impact modifier
    TSCA Listed: yes
    FDA Approval: CFR 178.3790
    NSF Approval: DWV only
    DSL Listing: yes
    European Approvals: Food approved in EEC for food contact
                        in any form
    Japanese Approvals: Japanese Positive List
    Form: white powder
    Specific Gravity: 1.09
    Bulk Density: 0.48 g/cc

Durastrength 200L:
    Main Features: excellent impact
                   low gloss
                   good weatherability
    Applications: siding capstock (low gloss)
                  siding substrate
                  profile (low gloss/reversion)
    Chemical Description: modified acrylic impact modifier
    TSCA Listed: yes
    FDA Approval: CFR 178.3790
    NSF Approval: no
    DSL Listing: yes
    European Approvals: Food approved in EEC for food contact
                        in any form
    Japanese Approvals: Japanese Positive List
    Form: white powder
    Specific Gravity: 1.09
    Bilk Density: 0.54 g/cc

Atofina Chemicals, Inc.: DURASTRENGTH High Efficiency Impact Modifiers (Continued):

Durastrength 300S
    Main Features: high performance
                  designed for lead stabilized systems
                  good weatherability
    Applications: profile (esp. Pb stabilized)
              injection molding
    Chemical Description: modified acrylic impact modifier
    TSCA Listed: yes
    FDA Approval: CFR 178.3790
    NSF Approval: no
    DSL Listing: yes
    European Approvals: Food approved in EEC for food contact in
                  any form
    Japanese Approvals: Japanese Positive List
    Form: white powder
    Specific Gravity: 1.09
    Bulk Density: 0.48 g/cc

Durastrength 500 Series:
    Main Features: excellent impact
                  tailored processability
                  good weatherability
    Applications: various
    Chemical Description: advanced acrylic modifiers
    TSCA Listed: yes
    FDA Approval: product specific
    NSF Approval: product specific
    DSL Listing: product specific
    European Approvals: product specific
    Japanese Approvals: product specific
    Form: white powder
    Specific Gravity: product specific
    Bulk Density: product specific

Mitsubishi Rayon America Inc.: METABLEN Impact Modifiers:

C-type: Impact Modifier (MBS)
    C-102
    C-132 (crease-whitening resistant type)
    C-140A (crease-whitening resistant type)
    C-320 C-201 C-201A C-202 C-350 C-301
    C-223A (Opaque type)
    C-323A (Opaque type for profile extrusion)

S-type: Silicone-acrylic weatherable impact modifier
    S-2001
    SRK-200 (High impact-resistance type for engineering plastics)

W-type: Acrylic weatherable impact modifier
    W-300A (High transparent type)

<----More excellent----
    AAA>AA>A>B>C

C-102:
    Izod Impact Strength: AA-A
    Failing Weight Impact Strength: AA-A
    Transparency: AA
    Crease whitening resistance: AA
    Hot water whitening resistance: AA
    Main applications: Sheets, bottles, general purpose

C-132:
    Izod impact strength: B
    Failing weight impact strength: A-B
    Transparency: AAA
    Crease whitening resistance: AAA
    Hot water whitening resistance: AAA
    Main applications: Films, sheets, high transparency and less
                      crease whitening

C-140A:
    Izod impact strength: A-B
    Failing weight impact strength: A
    Transparency: AAA-AA
    Crease whitening resistance: AAA-AA
    Hot water whitening resistance: AA
    Main applications: Films, sheets, high transparency and less
                      crease whitening

C-320:
    Izod impact strength: AA
    Failing weight impact strength: AA
    Transparency: AAA-AA
    Crease whitening resistance: AA-A
    Hot water whitening resistance: AA
    Main applications: Sheets, bottles, general purpose

Mitsubishi Rayon America Inc.: METABLEN Impact Modifiers (Continued):

<----More excellent---
     AAA>AA>A>B>C

C-201:
    Izod impact strength: AA
    Failing weight impact strength: AA
    Transparency: AA
    Crease whitening resistance: A
    Hot water whitening resistance: AA
    Main applications: Sheets, bottles, general purpose

C-201A:
    Izod impact strength: AA
    Failing weight impact strength: AA
    Transparency: AA
    Crease whitening resistance: A
    Hot water whitening resistance: AA
    Main applications: Sheets, bottles, general purpose

C-202:
    Izod Impact strength: AA
    Failing weight impact strength: AA
    Transparency: AA
    Crease whitening resistance: A
    Hot water whitening resistance: AA
    Main applications: Sheets, bottles, general purpose

C-350:
    Izod impact strength: AA
    Failing weight impact strength: AAA
    Transparency: AA-A
    Crease whitening resistance: A-B
    Hot water whitening resistance: A
    Main applications: Sheets, bottles, general purpose

C-301:
    Izod impact strength: AA
    Falling weight impact strength: AAA
    Transparency: A
    Crease whitening resistance: A-B
    Hot water whitening resistance: A
    Main applications: Sheets, bottles, general purpose

C-323A:
    Izod Impact Strength: AA
    Falling weight impact strength: AAA
    Hot water whitening resistance: A
    Main applications: Profile extrusion, injection, opaque

**Mitsubishi Rayon Co., Ltd.: METABLEN Impact Modifiers
(Continued):**

<----More excellent----
    AAA>AA>A>B>C

C-223A:
    Izod impact strength: AAA
    Falling weight impact strength: AA
    Hot water whitening resistance: A
    Main applications: Profile extrusion, injection, opaque

W-300A:
    Izod impact strength: B
    Falling weight impact strength: B
    Transparency: AA
    Crease whitening resistance: A
    Hot water whitening resistance: A
    Main applications: Sheets, corrugated sheets, weatherability,
                      transparency

S-2001:
    Izod impact strength: AAA-AA
    Falling weight impact strength: AA
    Main applications: Profile extrusion, engineering plastics,
                      opaque

SRK200:
    Izod impact strength: AA
    Falling weight impact strength: AA
    Main applications: Engineering plastics, opaque

Rohm and Haas Co.: MBS and Acrylic Impact Modifiers and Specialty Additives: Product Range:

MBS Tougheners:
EXL-2691/EXL-3691:
Features:
    Low Temperature Impact
    Thermal Stability: Fair*
    Colorability: Excellent
EXL-2691A/EXL-3691A:
Features:
    Low Temperature Impact
    Thermal Stability: Fair*
    Colorability: Excellent
    *For improved thermal stability, additional stabilizers must
     be added to the formulations

Acrylic Tougheners:
EXL-2330/EXL-3330:
Features:
    Weatherability
    Thermal Stability: Good
    Colorability: Fair
EXL-2335:
Features:
    Weatherability
    Thermal Stabiity: Good
    Colorability: Good
EXL-3361:
Features:
    Weatherability
    Thermal Stability: Excellent
    Colorability: Good

Specialty Additives:
EXL-5136:
Features:
    Weatherablity
    Thermal Stability: Good
    Colorability: Fair
EXL-5375:
Features:
    Weatherability
    Thermal Stability: Good
    Colorability: Fair
EXL-8619:
Features:
    Low Temperature Impact
    Thermal Stability: Excellent
    Colorability: N/A
    Clarity: Excellent

Section XI
Initiators

Aztec Peroxides, Inc.: Laporte, Plc: Catalysts & Initiators:

Accelerator A-305:
Polyester Curing: Accelerators
   Dimethyl aniline
   10% solution in phthalate

Accelerator C-101:
   Polyester Curing: Accelerators
   Cobalt octoate
   Solution in phthalate (1% cobalt)

Accelerator CA-12:
   Polyester Curing: Accelerator
   Cobalt/Amine
   Liquid mixture in phthalate

Accelerator COB-6:
   Polyester Curing: Accelerator
   Cobalt octoate
   Solution in xylene (6% cobalt)

Accelerator COB-10:
   Polyester Curing: Accelerators
   Cobalt octoate
   Solution in xylene (10% cobalt)

AAP-LA-M2:
   Polyester Curing: Ketone peroxides (Ambient temperature)
   Acetyl acetone peroxide
   Liquid mixture, low-activity

AAP-LA-M3:
   Polyester Curing: Ketone peroxides (Ambient temperature)
   Acetyl acetone peroxide
   Liquid mixture, low activity

AAP-NA-2:
   Polyester Curing: Ketone peroxides (Ambient temperature)
   Acetyl acetone peroxide
   Liquid mixture, normal mixture

APS:
   Persulfates
   Ammonium persulfate (ammonium peroxydisulfate)
   Technically pure salt

BCHPC:
   Polymerization: Peroxydicarbonates
   Bis(4-tert-butylcyclohexyl)peroxydicarbonate
   Powder, technically pure

Aztec Peroxides, Inc.: LaPorte, Plc: Catalysts & Initiators (Continued):

BCHPC:
    Polyester Curing: Peroxydicarbonates (50-100C)
    Bis(4-tert.butylcyclohexyl-peroxydicarbonate
    Powder, technically pure

BCHPC-40-SAQ:
    Polymerisation: Peroxydicarbonates
    Di(4-tert.butylcyclohexyl)peroxydicarbonate
    40% Suspension, aqueous

BCHPC-40-SAQ1:
    Polymerisation: Peroxydicarbonates
    Di(4-tert.butylcyclohexyl)peroxydicarbonate
    40% Aqueous suspension

BCHPC-75-W:
    Polymerisation: Peroxydicarbonates
    Bis(4-tert.butylcyclohexyl)peroxydicarbonate
    75% Powder, water damped

BCUP:
    Crosslinking: Aryl-Alkylperoxides
    Tert.-Butyl cumyl peroxide
    Liquid, techn. pure

BCUP-HA-M1:
    Polyester Curing: Alkyl-Arylperoxide (140-180C)
    Tert.Butylcumylperoxide
    Liquid mixture, high activity

BP-20-GY:
    Polyester Curing: Diacyl peroxides (Ambient temperature)
    Dibenzoyl peroxide
    20% Powder with gypsum

BP-20-SAQ:
    Polyester Curing: Diacyl peroxides (Ambient temperature)
    Dibenzoyl peroxide
    20% Aqueous suspension

BP-40-S:
    Polyester Curing: Diacyl peroxides (Ambient/Elevated tempera-
                    ture)
    Dibenzoyl peroxide
    40% suspension in phthalate

Aztec Peroxides, Inc.: LaPorte Plc: Catalysts & Initiators (Continued):

BP-40-SAQ:
   Dibenzoyl peroxide
   40% Aqueous suspension

BP-50-FT:
   Polyester Curing: Diacyl peroxides (Ambient/Elevated temperature)
   Dibenzoyl peroxide
   50% Powder with phthalate

BP-50-P1:
   Polyester Curing: Diacyl peroxides (Ambient temperature)
   Dibenzoyl peroxide
   50% Paste in phthalate

BP-75-W:
   Polymerisation: Diacyl peroxides
   Dibenzoyl peroxide
   75% Powder, water damped

BP-75-W 7:
   Dibenzoyl peroxide
   Powder, water damped

BU-50-AL:
   Polymerisation: Perketals
   2,2-Di(tert.butylperoxy)butane
   50% Solution in aliphatics

BU-50-WO:
   Polymerisation: Perketals
   2,2-Di(tert-butylperoxy)butane
   50%, Solution in white oil

CCDFB:
   Polymerisation: C-C Initiators
   2,3-Dimethyl-2,3-diphenyl butane
   Flakes, technically pure

CCDFB-90:
   Polymerisation: C-C Initiators
   2,3-Dimethyl-2,3-diphenyl butane
   Flakes, technically pure

CCDFH:
   Polymerisation: Other Initiators
   3,4-Dimethyl-3,4-diphenyl hexane
   Viscous liquid

Aztec Peroxides, Inc.: LaPorte, Plc: Catalysts & Initiators (Continued):

CCPIB:
   Polymerisation: Other initiators
   Poly 1,4-diisopropyl benzene
   Flakes, technically pure

CEPC:
   Polymerisation: Peroxydicarbonates
   Dicetylperoxydicarbonate
   Flakes, technically pure

CEPC-40-SAQ:
   Polymerisation: Peroxydicarbonates
   Dicetylperoxydicarbonate
   40% Suspension, aqueous

CEPC-40-SAQ1:
   Polymerisation: Peroxydicarbonates
   Dicetylperoxydicarbonate
   40%, Aqueous suspension

CEPC-50-SAQ1:
   Polymerisation: Peroxydicarbonates
   Dicetylperoxydicarbonate
   50%, Aqueous suspension

CH-25-WO:
   Polymerisation: Perketals
   1,2-Di(tert.butylperoxy)cyclohexane
   25% Solution in white oil

CH-50-AL:
   Polymerisation: Perketals
   1,1-Di(tert.-butylperoxy)cyclohexane
   50% Solution in aliphatics

CH-50-AL:
   Polyester Curing: Perketals (120-160C)
   1,1-Di(tert.-butylperoxy)cyclohexane
   50% Solution in aliphatics

CH-50-FT:
   Polyester Curing: Perketals (120-160C)
   1,2-Di(tert.-butylperoxy)cyclohexane
   50% Solution in phthalic acid ester

CH-50-WO:
   Polymerisation: Perketals
   1,1-Di(tert.butylperoxy)cyclohexane
   50% Solution in white oil

Aztec Peroxides, Inc.: LaPorte, Plc: Catalysts & Initiators (Continued):

CHP-90-W1:
   Polyester Curing: Ketone peroxides (Ambient temperature)
   Cyclohexanone peroxide
   90% powder, water damped

CH-90-W2:
   Polyester Curing: Ketone peroxides (Ambient temperature)
   Cyclohexanone peroxide
   90% powder, water damped

CHPC:
   Polymerisation: Peroxydicarbonates
   Dicyclohexyl-peroxydicarbonate
   Powder, technically pure

CHPC-40-SAQ1:
   Polymerisation: Peroxydicarbonates
   Dicyclohexylperoxydicarbonate
   40%, Aqueous suspension

CHPC-90-W:
   Polymerisation: Peroxydicarbonates
   Dicyclohexyl-peroxydicarbonate
   90% Powder, water damped

CHP-HA-M2:
   Polyester Curing: Ketone peroxides (Ambient temperature)
   Cyclohexanone peroxide
   Liquid mixture, high activity

CHP-HA-M4D:
   Polyester Curing: Ketone Peroxides (Ambient temperature)
   Cyclohexanone peroxide
   Liquid mixture, high activity

CHP-NA-1:
   Polyester Curing: Ketone peroxides (Ambient temperature)
   Cyclohexanone peroxide
   Solution, normal activity

CHP-SA-1:
   Polyester Curing: Ketone peroxides (Ambient temperature)
   Cyclohexanone peroxide
   Solution, super activity

CUHP-80:
   Polymerisation: Hydroperoxide
   Cumene hydroperoxide
   80% Solution in cumene

Aztec Peroxides, Inc.: LaPorte, Plc: Catalysts & Initiators (Continued):

CUHP-90:
    Polymerisation: Hydroperoxide
    Cumene hydroperoxide
    89% Solution in cumene

CUPND-75-AL1:
    Polymerisation: Peresters
    Cumylperneodecanoate
    75% Solution in Aliphatics

CUROX:
    Persulfates
    Potassium monosulfate (KMPS)
    Potassium Peroxymonosulfate-triple salt: 2KHSO5-KHSO4-K2SO4

DCLBP-50-PSI:
    Crosslinking: Diacylperoxides
    Di(2,4-dichlorobenzoyl)peroxide
    50% Paste in silicone oil

DCUP:
    Crosslinking: Aryl-Alkylperoxides
    Dicumylperoxide
    Powder, techn. pure

DCUP-1:
    Crosslinking: Aryl-Alkylperoxides
    Dicumylperoxide
    Powder, techn. pure

DCUP-40-IC:
    Polyester Curing: Aryl-Alkylperoxides (130-170C)
    Dicumylperoxide
    40%, powder with chalk

DCUP-40-IC:
    Crosslinking at above 160C
    Dicumylperoxide
    40%, Powder with chalk

DCUP-40 IC 1:
    Crosslinking at above 160C
    Dicumylperoxide
    40%, Powder with kaolin

DHBP:
    Crosslinking: Dialkylperoxides
    2,5-Dimethyl 2,5-di(tert.butyl peroxy) hexane
    Liquid, techn. pure

Aztec Peroxides, Inc.:Laporte, Plc: Catalysts & Initiators (Continued):

DHBP-45-IC:
   Crosslinking: Dialkylperoxides
   2,5-Dimethyl-2,5-di(tert.butylperoxy)hexane
   45%, powder

DHBP-20-IC5:
   Crosslinking: Dialkyl peroxides
   2,5-Dimethyl-2,5-di(tert.butylperoxy)hexane
   20%, granules

DHBP-45-IC 1:
   Crosslinking: Dialkylperoxides
   2,5-Dimethyl-2,5-di(tert.butylperoxy)hexane
   45%, Powder

DHBP-45-PSI:
   Crosslinking: Dialkylperoxides
   2,5-Dimethyl-2,5-di(tert.butylperoxy)hexane
   45%, Paste in silicone rubber

DHBP-7.5-IC5:
   Crosslinking: Dialkylperoxides
   2,5-Dimethyl-2,5-di(tert.butylperoxy)hexane
   7.5%, Granules

DHBP-75-PIC:
   Crosslinking: Dialkyl peroxides
   2,5-Dimethyl-2,5-di(tert.butylperoxy)hexane
   72%, paste in silicagel

DHBP-75-SR:
   Crosslinking: Dialkylperoxides
   2,5-Dimethyl 2,5-di(tert.butyl peroxy) hexane
   Liquid, 75%, solution in inhibitor

DHPEH:
   Polyester Curing: Peresters (70-130C)
   2,5-Dimethyl-2,5-di(2-ethylhexanoyl-peroxy)hexane
   Liquid, technically pure

DHPEH:
   Polymerisation: Peresters
   2,5-Dimethyl-2,5-di(2-ethylhexanoyl-peroxy)hexane
   Liquid, technically pure

DIPND-50-AL:
   Polymerisation: Peresters
   1,3-Di(2-neodecanoylperoxy isopropyl)benzene
   50%, solution in aliphatics

Aztec Peroxides, Inc.: Laporte, Plc: Catalysts & Initiators (Continued):

DIPP-2:
    Crosslinking at above 170C
    1,2-Di(2-tert.butylperoxy isopropyl)benzene
    technically pure, flakes

DIPP-2:
    Polyester Curing: Aryl-Alkyl peroxides (140-200C)
    1,3-Di(2-tert.butylperoxy isopropyl) benzene
    Flakes, technically pure

DIPP-40-IC:
    Crosslinking at above 170C
    1,3-Di(2-tert.butylperoxy isopropyl)benzene
    40%, powder with chalk

DIPP-40-IC5:
    Crosslinking: Dialkylperoxides
    Di(tert.butylperoxy isopropyl)benzene
    40%, Powder with Polypropylene

DP:
    Polymerisation: Diacylperoxides
    Didecanoylperoxide
    Flakes, technically pure

DTAP:
    Polymerisation: Aryl-Alkylperoxides
    Di(tert.amyl)peroxide
    Liquid, techn. pure

DTBP:
    Polymerisation: Aryl-Alkylperoxides
    Di(tert.butyl)peroxide
    Liquid, techn. pure

DTBP
    Crosslinking: Dialkylperoxides
    Di(tert.butyl) peroxide
    Liquid, techn. pure

DTBP-30-AL:
    Polymerisation: Aryl-Alkylperoxides
    Di(tert.butyl)peroxide
    30%, solution in aliphatics

DTBP-30-AL-2:
    Polymerisation: Aryl-Alkylperoxides
    Di(tert.butyl) peroxide
    30%, solution in aliphatics (Isopar E)

Aztec Peroxides, Inc.: Laporte, Plc: Catalysts & Initiators (Continued):

DYBP:
    Crosslinking: Dialkyl Peroxides
    2,5-Dimethyl 2,5-di(tert.butylperoxy)hexine-3
    Technically pure, liquid

DYBP-85-WO:
    Crosslinking: Dialkylperoxides
    2,5-Dimethyl 2,5-di(tert.butylperoxy)hexine-3
    85% solution in white oil

EHPC-40-EAQ:
    Polymerisation: Peroxydicarbonates
    Di-2-ethylhexyl peroxydicarbonate
    40%, aqueous emulsion (frozen flakes)

EHPC-40-ENF:
    Polymerisation: Peroxydicarbonates
    Di-2-ethylhexyl peroxydicarbonate
    40%, non freezing emulsion

EHPC-50-EAQ:
    Polymerisation: Peroxydicarbonates
    Di-2-ethylhexyl peroxydicarbonate
    50% Aqueous Emulsion (frozen flakes)

EHPC-50-NF:
    Polymerisation: Peroxydicarbonates
    Di-2-ethylhexyl peroxydicarbonate
    50% non freezing emulsion

EHPC-65-AL:
    Polymerisation: Peroxydicarbonates
    Di-2-ethylhexyl peroxydicarbonate
    65%, solution in aliphatics

EHPC-75-AL:
    Polymerisation: Peroxydicarbonates
    Di-2-ethylhexyl peroxydicarbonate
    75%, Solution in aliphatics

Inhibitor BC-500:
    Polyester Curing Inhibitors
    Di-tert.butyl p-cresol
    40% Solution in xylene

Aztec Peroxides, Inc.: Laporte Plc: Catalysts & Initiators (Continued):

Inhibitor TC-510:
    Polyester Curing: Inhibitors
    tert.Butyl catechol
    10% Solution in styrene

INP-37-AL:
    Polymerisation: Diacylperoxides
    Di(3,5,5-trimethyl-hexanoyl)peroxide
    37%, Solution in aliphatics

INP-40-ENF:
    Polymerisation: Diacylperoxides
    Di(3,5,5-trimethyl-hexanoyl)peroxide
    40%, Emulsion non-freezing

INP-75-AL:
    Polymerisation: Diacylperoxides
    Di(3,5,5-trimethyl-hexanoyl)peroxide
    75%, Solution in aliphatics

KMPS:
    Persulfates
    Potassium monopersulfate (KMPS)
    Potassium peroxymonosulfate-triple salt: 2KHSO5-KHSO4-K2SO4

KPS:
    Persulfates
    Potassium persulfate (potassium peroxodisulfate)
    technically pure, salt

LP:
    Polymerisation: Diacylperoxides
    Dilauroyl peroxide
    Flakes, technically pure

LP-25-SAQ:
    Polymerisation: Diacyl peroxides
    Dilauroyl peroxide
    25% aqueous suspension

LP-40-SAQ:
    Polymerisation: Diacyl peroxides
    Dilauroyl peroxide
    40% suspension, aqueous

Aztec Peroxides, Inc.: Laporte Plc: Catalysts & Initiators (Continued):

MEKP-HA-2:
   Polyester Curing: Ketone peroxides (Ambient temperature)
   Methyl ethyl ketone peroxide
   Liquid mixture, high activity

MEKP-LA-3:
   Polyester Curing: Ketone peroxides (Ambient temperature)
   Methyl ethyl ketone peroxide
   Liquid mixture, low activity

MEKP-NA-M3:
   Polyester Curing: Ketone peroxides (Ambient temperature)
   Methyl ethyl ketone peroxide
   Liquid mixture, normal activity

MEKP-S:
   Polyester Curing: Ketone peroxides (Ambient temperature)
   Methyl ethyl ketone peroxide
   Liquid mixture, high activity

MEKP-SA-1:
   Polyester Curing: Ketone peroxides (Ambient temperature)
   Methyl ethyl ketone peroxide
   Liquid mixture, super activity

MIKP-LA-M1:
   Polyester Curing: Ketone peroxides (60-120C)
   Methyl isobutyl ketone peroxide
   Liquid mixture, low activity

MIKP-NA-1:
   Polyester Curing: Ketone peroxides (60-120C)
   Methyl isobutyl ketone peroxide
   Liquid mixture, normal activity

MIKP-NA-4:
   Polymerisation: Ketone peroxides
   Methyl isobutyl ketone peroxide
   Liquid mixture, normal activity

MIKP-NA-M2:
   Polyester Curing: Ketone Peroxides (60-120C)
   Methyl Isobutyl ketone peroxide
   Liquid mixture, normal activity

Aztec Peroxides, Inc.: Laporte Plc: Catalysts & Initiators (Continued):

MYPC:
Polymerisation: Peroxydicarbonates
Dimyristylperoxydicarbonate
Flakes, technically pure

MYPC-40-SAQ:
Polymerisation: Peroxydicarbonates
Dimyristylperoxydicarbonate
40% Suspension, aqueous

NPS:
Sodium persulfate (sodium peroxodisulfate)
Technically pure, salt

PMB-50-PSI:
Crosslinking: Diacylperoxides
Di(4-methylbenzoyl)peroxide
50%, Paste in silicone oil

SBPC-60-AL1:
Polymerisation: Peroxydicarbonates
sec-Butyl Peroxydicarbonate
60%, Solution in aliphatics

SUCP-70-W:
Polymerisation: Diacylperoxides
Disuccinic acid peroxide
70% powder, water damped

TAHP-80-AL:
Polymerisation: Hydroperoxides
tert.Amyl hydroperoxide
80%, Solution in aliphatics

TAPB:
Polymerisation: Diacylperoxides
tert.-Amylperoxy benzoate
Liquid, technical pure

TAPEH:
Polymerisation: Peresters
tert.Amylperoxy-2-ethyl hexanoate
Liquid, technically pure

TAPND-75-AL:
Polymerisation: Peresters
tert.Amylperneodecanoate

75%, Solution in aliphatics

Aztec Peroxides, Inc.: Laporte Plc: Catalysts & Initiators (Continued):

TAPPI-30-AL:
Polymerisation: Peresters
tert.Amylperoxy pivalate
30%, Solution in aliphatics

TAPPI-75-AL:
Polymerisation: Peresters
tert.Amylperoxy pivalate
75%, Solution in aliphatics

TBHP-70-AQ:
Polymerisation: Hydroperoxides
tert.Butyl-hydroperoxide
70%, Aqueous solution

TBHP-78-AQ:
Polymerisation: Hydroperoxides
tert.Butyl-hydroperoxide
78%, Aqueous solution

TBHP-80:
Polymerisation: Hydroperoxides
tert.Butyl-hydroperoxide
80%, liquid

TBPA-30-AL2:
Polymerisation: Peresters
tert.Butylperoxyacetate
30% solution in aliphatics (Isopar E)

TBPA-50-AL:
Polymerisation: Peresters
tert.Butylperoxyacetate
50%, Solution in aliphatics

TBPB:
Polymerisation: Peresters
tert.Butylperoxy benzoate
Liquid, techn.pure

TBPB-50-FT:
Polymerisation: Peresters
tert.Butylperoxy benzoate
50% Solution in phthalic acid ester

**Aztec Peroxides: LaPorte PLC: Catalysts and Initiators
(Continued):**

TBPB-75-AL:
    Polymerisation: Peresters
    tert.Butylperoxy benzoate
    75%, solution in aliphatics

TBPEH:
    Polymerisation: Peresters
    tert.Butyl-per-2-ethyl hexanoate
    Liquid, techn.pure

TBPEH-30-AL:
    Polymerisation: Peresters
    tert.Butylperoxy-2-ethyl hexanoate
    30%, solution in aliphatics

TBPEH-30-AL-2:
    Polymerisation: Peresters
    tert.Butyl-per-2-ethyl hexanoate
    30%, solution in aliphatics (Isopar E)

TBPEH-50-AL:
    Polymerisation: Peresters
    tert. butyl-peroxy-2-ethyl hexanoate
    50%, solution in aliphatics

TBPEH-70-AL:
    Polymerisation: Peresters
    tert.Butylperoxy-2-ethyl hexanoate
    70%, solution in aliphatics

TBPEHC:
    Polymerisation: Perester
    tert.Butylperoxy-(2-ethylhexyl)carbonate
    Liquid, technically pure

TBPIC-75-AL:
    Polymerisation: Peresters
    tert.Butylperoxy-isopropylcarbonate
    75%, Solution in aliphatics

TBPIN:
    Polymerisation: Peresters
    tert.Butylperoxy-3,5,5-trimethylhexanoate
    Liquid, technically pure

**Aztec Peroxides, Inc.: Laporte Plc: Catalysts & Initiators
(Continued):**

TBPIN-30-AL:
    Polymerisation: Peresters
    tert.Butylperoxy-3,5,5-trimethyl hexanoate
    30%, solution in aliphatics

TBPND:
    Polymerisation: Peresters
    tert. Butylperneodeacanoate
    Liquid, technically pure

TBPND-75-AL:
    Polymerisation: Peresters
    tert.Butylperneodecanoate
    75% solution in aliphatics

TBPPI-25-AL:
    Polymerisation: Peresters
    tert.Butylperpivalate
    25% Solution in aliphatics

TBPPI-25-AL2:
    Polymerisation: Peresters
    tert. Butylperpivalate
    25% Solution in aliphatics (Isopar E)

TBPPI-75-AL:
    Polymerisation: Peresters
    tert.Butylperpivalate
    75% Solution in aliphatics

TMCH-40-IC:
    Crosslinking: Perketals
    1,1-Di(tert.butylperoxy)-3,3,5-trimethyl cyclohexane
    40%, Powder with chalk

TMCH-50-AL:
    Polymerisation: Perketals
    1,1-Bis(tert.butylperoxy)3,3,5-trimethyl cyclohexane
    50% Solution in aliphatics

TMCH-50-FT:
    Polymerisation: Perketals
    1,1-Di(tert.butylperoxy)3,3,5-trimethyl-cyclohexane
    50%, Solution in phthalate

TMCH-75-AL:
    Polymerisation: Perketals
    1,1-Bis(tert.butylperoxy)-3,3,5-trimethyl cyclohexane
    75% Solution in aliphatics

Sartomer Co.: ESACURE Photoinitiators:

Esacure EB 3:
   Photoinitiator Esacure EB 3 is a stabilized liquid mixture
of benzoin normal butyl ethers, which offers easy processibility
even at low temperatures to a variety of UV cure systems.
   Clear, yellow, oily liquid
   Odor: Faint, characteristic
   Freezing point, C: <5
   Density, g/ml: 1.060

Esacure KB 1:
   Photoinitiator Esacure KB 1 is a crystalline powder of
benzildimethyl ketal, which is readily soluble in most organic
solvents, monomers and photoreactive polymers. The optimum
absorption is at 250-350 nm. It has fast cure response and
good through cure.
   Supplied form: 100% solid
   Appearance: White crystalline powder
   Odor: Faint, characteristic
   Molecular weight: 256.30

Esacure KIP-100F:
   A liquid mixture of an oligomeric alpha hydroxy ketone and
2-hydroxy-2-methyl-1-phenyl 1-propanone. It is a highly reactive,
non-yellowing initiator for the photopolymerization of UV
curable systems based on an acrylic unsaturated oligomers and
monomers.
   Clear, slightly yellow viscous liquid
   Odor: Faint, characteristic
   Boiling Point: >200C

Esacure KIP-150:
   A polymeric hydroxy ketone. It is a highly reactive, non-
yellowing initiator for the photopolymerization of U.V.
curable systems based on acrylic unsaturated oligomers and
monomers.
   Orange-yellow semi-solid
   Odor: Faint, characteristic
   Color: Max 13 Gardner
   Boiling Point: >300C

Sartomer Co.: ESACURE Photoinitiators (Continued):

Esacure KT 37:
   A liquid mixture of two photoinitiators, oligomeric KIP 150F
and TZT. It is a highly reactive mixture of photoinitiators,
particularly designed for the photopolymerization of UV light
curable systems based on acrylic and methacrylic unsaturated
oligomers and monomers. The radicals' generation occurs via both
splitting and hydrogen abstraction mechanisms. Esacure KT 37
is very effective in combination with tertiary amine type
coinitiators, which overcome oxygen inhibition, allowing fast
cure rates.
   Clear, slightly yellow liquid
   Odor: Faint, characteristic
   Boiling point: 310C at 760 mm Hg

Esacure KTO46:
   A blend of 2,4,6-Trimethylbenzoyldiphenylphosphine oxide,
alpha-hydroxyketone and benzophenone derivative. It is a liquid
photoinitiator which can be easily incorporated by simply
stirring into your resin system.
   Clear, slightly yellow liquid
   Odor: Faint, characteristic
   Flash Point: 118C
   Density: 1.106+-0.005 g/ml at 20C

Esacure TZT:
   TZT is clear, low viscosity liquid with optimum absorbance
at 250-300 nm. It is highly soluble and highly reactive with
a flash point of 140C.
Formulation Properties:
   Flexibility   Hardness
   Abrasion Resistance   Water Resiatance
   Low Odor   Lower Volatility
   Clear, colorless to light straw liquid
   Odor: Very faint, characteristic
   Pour Point: -7C
   Boiling Range: 310-330C at 760 mm Hg

Esacure X 33:
   A blend of components noted, which enables generation of free
radicals by the hydrogen-abstraction mechanism, giving fast
surface cure speed. Esacure X33 is a liquid photoinitiator
which can be easily incorporated by simply stirring into your
resin system.
   Clear, brown liquid
   Odor: Faint, characteristic
   Color: Max 13 Gardner
   Flash Point: 140C
   Density: 1.095+-0.02 g/ml at 20C

Sartomer Co.: Photosensitizer/Coinitiators:

Benzophenone:
A white to off-white crystalline solid. It is also available in flaked form.
Benzophenones are photosensitizers for use in UV curing systems such as printing inks and wood and metal finishes.
Purity: 99.0% min.
Melt range: 48-49C
Molecular Weight: 182.21
CAS No.: 119-61-9

CN-383 Reactive Amine Coinitiator:
A monofunctional amine synergist which, when used in conjunction with a photosensitizer such as benzophenone, promotes rapid curing under UV light. Additional benefits include reduced odors both at press side and in the cured film, reduced blooming or migration, and color fading of selective pigments such as reflex blue and rhodamine red.
Performance Properties:
Adhesion
Hardness
Appearance: Slightly Hazy
Color, APHA: 225
Specific Gravity @ 25C: 0.9903
Viscosity @ 25C, cps: 18
Total Amine (mg KOH/g): 135

CN-384 Reactive Amine Coinitiator:
A difunctional amine synergist which, when used in conjunction with a photosensitizer such as benzophenone, promotes rapid curing under UV light. Additional benefits include reduced odors both at press side and in the cured film, reduced blooming or migration, and color fading of selective pigments such as reflex blue and rhodamine red.
Performance Properties:
Adhesion
Hardness
Appearance: Clear
Color, APHA: 100
Specific Gravity @ 25C: 1.0534
Viscosity @ 25C, cps: 115
Total Amine (mg KOH/g): 103

Sartomer Co.: SARCAT Cationic Photoinitiators:

SarCat CD-1010 Cationic Photoinitiator:
CD-1010 is a 50/50 blend of triaryl sulfonium hexafluoro-antimonate in propylene carbonate. This cationic photoinitiator offers fast cure speeds in epoxy, vinyl ether and other cationically cured resin systems.
Suggested Applications:
CD-1010 is recommended for use in metal, paper, wood and plastic coatings and inks. Cationically cured coatings provide enhanced adhesion particularly to metallic substrates.
Appearance: Clear Amber Liquid
Solids, %: 48-52
Viscosity, 25C, cps: 60-90
Weight, lbs/gal: 11.7
Color, Gardner: 3-6

SarCat CD-1011 Cationic Photoinitiator:
A 50/50 blend of triaryl sulfonium hexafluorophosphate in propylene carbonate. This cationic photoinitiator offers fast cure speeds in epoxy, vinyl ether and other cationically cured resin systems.
Suggested Applications:
CD-1011 is recommended for use in metal, paper, wood and plastic coatings and inks. Cationically cured coatings provide enhanced adhesion particularly to metallic substrates.
Appearance: Clear pale yellow liquid
Solids, %: 48-52
Viscosity @ 25C, cps: 60-90
Weight, lbs/gal: 11.0
Color, Gardner: 6

SarCat CD-1012 Cationic Photoinitiator:
CD-1012 is a diaryl iodonium hexafluoroantimonate. This cationic photoinitiator offers fast cure speeds in epoxy, vinyl ether and other cationically cured resin systems. It has better solubility than similar iodonium salts.
Suggested Applications:
CD-1012 is recommended for use in plastic, paper, wood and metal coatings and inks. Cationically cured coatings provide enhanced adhesion particularly to metallic substrates.
Appearance: Off-white powder
Melting point, C: 90-92
Solids, %: 100

Witco Corp.: Additives for Vinyl: Organic Peroxides:

As a leading manufacturer and developer of organic peroxides, Witco has produced many highly effective products that have become industry standards. And to keep pace with customers' needs, these development efforts are ongoing.

Organic peroxides act through the splitting of the --O--O-- bond into free radicals, thereby initiating the polymerization or crosslinking of monomers or polymers. Their exceptionally broad line includes diacyl peroxides, dialkyl peroxides, hydroperoxides, ketone peroxides, peroxyketals, peroxydicarbonates, and peroxyesters. The last two are particularly important in PVC resin manufacture as initiators in the polymerization of vinyl chloride monomer.

Comparative Polymerizations of Vinyl Chloride
Conversion Rate at 55C

Initiator:
ESPEROX 551M (t-amyl peroxypivalate):
  % Wt.: 0.107
  % Conversion: 1.5 hrs.: 12.0
               3.5 hrs.: 44.5
               5.5 hrs.: 76.5

Esperox 31M (t-butyl peroxypivalate):
  % Wt.: 0.107
  % Conversion: 1.5 hrs.: 11.5
               3.5 hrs.: 42.0
               5.5 hrs.: 77.5

Esperox 750M (t-butyl peroxyneoheptanoate):
  % Wt.: 0.067
  % Conversion: 1.5 hrs.: 17.5
               3.5 hrs.: 48.5
               5.5 hrs.: 81.5

ESPERCARB 840M (di-2-ethyl hexyl peroxydicarbonate):
  % Wt.: 0.065
  % Conversion: 1.5 hrs.: 19.5
               3.5 hrs.: 57.0
               5.5 hrs.: 85.0

Esperox 33M (t-butyl peroxy neodecanoate):
  % Wt.: 0.065
  % Conversion: 1.5 hrs.: 19.0
               3.5 hrs.: 49.5
               5.5 hrs.: 80.5

Esperox 939 (cumyl peroxy neodecanoate):
  % Wt.: 0.113
  % Conversion: 1.5 hrs.: 36.0
               3.5 hrs.: 71.5
               5.5 hrs.: 80.5

Section XII
Lubricants

Acme-Hardesty Co.: JENKINOL Lubricants for Plastics:

Application: PVC Flexible Calendering:
Jenkinol L 240:
    Functionality: Internal   Use Level: 0.8-1.5 phr
    Comments: Reduces compound viscosity, aids bank distribution
    and reduces flow lines. Replaces up to 50% of process aid.
Jenkinol L 230*:
    Functionality: External   Use Level: 0.2-0.5 phr
    Comments: Excellent for Pb and Ca/Zn formulations
Jenkinol L 221:
    Functionality: External   Use Level: 0.1-0.3 phr
    Comments: For rigid and flexible PVC formulations.
Jenkinol L 273:
    Functionality: Internal/External   Use Level: 0.5-1.2 phr
    Comments: Functionality dependent upon usage
Jenkinol 680:
    Functionality: Co-stabilizer
    Comments: Epoxidized soybean oil (ESO)
Jenkinol L 271:
    Functionality: External   Use Level: 0.3-0.8 phr
    Comments: Good for all mixed metal stabilizer systems

Application: PVC Rigid Calendering:
Jenkinol L 210*:
    Functionality: Internal   Use Level: 0.8-1.2 phr
    Comments: Will not affect product clarity
Jenkinol L 216*:
    Functionality: Internal   Use Level: 0.8-1.5 phr
    Comments: Plate-out free. Will not affect product clarity.
Jenkinol L 260:
    Functionality: Internal   Use Level: 0.5-1.2 phr
    Comments: Reduces die swell and internal compound viscosity.
Jenkinol 680:
    Functionality: Co-stabilizer
    Comments: Epoxidized soybean oil (ESO).
Jenkinol L 271:
    Functionality: External   Use Level: 0.3-0.8 phr
    Comments: For semirigid and flexible PVC compounds. Release
          agent in injection molded polyurethane.
Jenkinol L 276*:
    Functionality: External   Use Level: 0.3-0.8 phr
    Comments: For semirigid and flexible PVC compounds. Release
          agent in injection molded polyurethane.

*FDA Approved

Acme-Hardesty Co.: JENKINOL Lubricants for Plastics (Continued):

Application: PVC Extrusion Blow Molding:
Jenkinol L 210*:
    Functionality: Internal   Use Level: 1.0-1.5 phr
    Comments: Will not affect clarity.
Jenkinol L 212*:
    Functionality: Internal   Use Level: 1.0-1.5 phr
    Comments: Viscosity depressant in ejection molded compounds.
Jenkinol L 215*:
    Functionality: Internal   Use Level: 0.8-1.5 phr
    Comments: Good for Ca/Zn stabilizer systems.
Jenkinol L 319:
    Functionality: External   Use Level: 0.5-0.8 phr
    Comments: Release agent for polycarbonate compounds.
Jenkinol L 321*:
    Functionality: Internal   Use Level: 0.8-1.5 phr
    Comments: Will not affect clarity. Boosts thermal stability
              in Ca/Zn formulations.
Jenkinol L 729:
    Functionality: External   Use Level: 0.05-0.2 phr
    Comments: Oxidized PE wax.
Jenkinol L 276*:
    Functionality: External   Use Level: 0.3-0.8 phr
    Comments: For semirigid and flexible PVC compounds. Release
              agent in injection molded polyurethane.
Jenkinol L 280*:
    Functionality: External   Use Level: 0.8-1.5 phr
    Comments: Mold release agent.

Application: PVC Rigid Extrusion:
Calcium Stearate:
    Functionality: Internal/External   Use Level: 0.5-1.5 phr
    Comments: Functionality dependent upon use level.
Jenkinol L 212*:
    Functionality: Internal   Use Level: 0.5-0.8 phr
    Comments: Internal viscosity depressant.
Jenkinol L 216*:
    Functionality: Internal   Use Level: 0.6-1.2 phr
    Comments: Will not plate out or affect clarity.
Jenkinol L 221:
    Functionality: External   Use Level: 0.1-0.3 phr
    Comments: Aids in pigment/filler dispersion.
Jenkinol L 230*:
    Functionality: External   Use Level: 0.2-0.5 phr
    Comments: Used in vinyl formulations which are tin, lead
              and Ca/Zn stabilized
Jenkinol L 232*:
    Functionality: Internal/External   Use Level: 1.2 phr maximum
    Comments: Functionality dependent upon use level.
Jenkinol L 260:
    Functionality: Internal   Use Level: 0.6-1.5 phr
    Comments: Reduces die swell and internal compound viscosity.

*FDA Approved

Acme-Hardesty Co.: JENKINOL Lubricants for Plastics (Continued):

Application: PVC Rigid Extrusion (Continued):
Jenkinol L 265:
    Functionality: Internal   Use Level: 0.5-2.5 phr
    Comments: Ideal for high flow requirements.
Jenkinol L 270:
    Functionality: External   Use Level: 0.4-0.7 phr
    Comments: Additive in chlorinated PVC
Jenkinol L 276*:
    Functionality: External   Use Level: 0.3-0.8 phr
    Comments: For semirigid and flexible PVC compounds. Release
           agent in ejection molded polyurethene
Jenkinol L 280*:
    Functionality: External   Use Level: 0.8-1.5 phr
    Comments: For rigid PVC compounds.
Jenkinol L 292:
    Functionality: Internal   Use Level: 0.8-1.5 phr
    Comments: For non-clear applications only.
Jenkinol L 319:
    Functionality: External   Use Level: 0.5-0.8 phr
    Comments: Will not plate-out or affect clarity.
Jenkinol L 321*:
    Functionality: Internal   Use Level: 0.5-1.3 phr
    Comments: Controls fusion and die swell.
Jenkinol L 207:
    Functionality: External   Use Level: 0.3-0.9 phr
    Comments: Non-oxidized PE wax. Improved efficiency over
           paraffin waxes.
Jenkinol L 729:
    Functionality: External   Use Level: 0.05-0.2 phr
    Comments: Oxidized PE wax.

Application: PVC Injection Molding:
Jenkinol L 212:
    Functionality: Internal   Use Level: 0.5-1.5 phr
    Comments: Viscosity depressant.
Jenkinol L 221:
    Functionality: Internal   Use Level: 0.1-0.3 phr
    Comments: Pb stabilized systems.
Jenkinol L 260:
    Functionality: Internal   Use Level: 0.6-1.5 phr
    Comments: Reduces die swell and internal compound viscosity
Jenkinol L 265:
    Functionality: Internal   Use Level: 0.5-2.5 phr
    Comments: Ideal for high flow requirements.
Jenkinol L 276*:
    Functionality: External   Use Level: 0.5-0.8 phr
    Comments: For semirigid and flexible PVC compounds. Release
           agent in ejection molded polyurethene.

*FDA Approved

Acme-Hardesty Co.: JENKINOL Lubricants for Plastics (Continued):

Application: PVC Injection Molding (Continued):
Jenkinol L 292:
    Functionality: Internal   Use Level: 0.1-0.2 phr
    Comments: Non-clear applications.
Jenkinol L 319:
    Functionality: External   Use Level: 0.5-0.8 phr
    Comments: No plate-out.

Application: PVC Blown Film:
Jenkinol L 215*:
    Functionality: Internal   Use Level: 0.8-1.5 phr
    Comments: No affect on clarity.
Jenkinol L 216*:
    Functionality: Internal   Use Level: 1.2-1.5 phr
    Comments: Low free glyceride and reduced risk of plate-out.
Jenkinol L 321*:
    Functionality: Internal   Use Level: 0.5-1.3 phr
    Comments: Controls fusion and die swell.
Jenkinol 680:
    Functionality: Co-stabilizer
    Comments: Epoxidized soybean oil (ESO).
Jenkinol L 270:
    Functionality: External   Use Level: 0.4-0.7 phr
    Comments: Improves anti-block.
Jenkinol L 276*:
    Functionality: External   Use Level: 0.3-0.8 phr
    Comments: Improves anti-block.

Application: Polypropylene:
Calcium Stearate:
    Functionality: Chain length modifier Use Level: 40-60 ppm
    Comments: Heat stable
Ethylene Bis Stearamide (EBS):
    Functionality: Antistat   Use Level: 1-4 phr
    Comments: Powder and beads available
Zinc Stearate:
    Functionality: Chain length modifier Use Level: 40-60 ppm
    Comments: Heat stable

Application: Polyethylene:
Calcium Stearate:
    Functionality: Chain length modifier Use Level: 40-60 ppm
    Comments: Heat stable

*FDA Approved

Acme-Hardesty Co.: JENKINOL Lubricants for Plastics (Continued):

Application: Polystyrene:
Zinc Stearate:
    Functionality: Chain Length Modifier   Use Level: 40-60 ppm
    Comments: Heat Stable
Magnesium Stearate:
    Functionality: Chain Length Modifier   Use Level: 40-60 ppm
    Comments: Heat Stable
Aluminum Stearate:
    Functionality: Chain Length Modifier   Use Level: 40-60 ppm
    Comments: Heat Stable

Application: Polycarbonate:
Jenkinol L 319:
    Functionality: External   Use Level: 0.5-0.8 phr
    Comments: Will not plate-out or affect clarity
Jenkinol L 232*:
    Functionality: Internal/External   Use Level: 1.2 phr max
    Comments: Will lower polycarbonate viscosity recommended
            for use with regrind.

Application: Polyurethane:
Urethane-Grade Castor Oil:
    Functionality: Polyol   Use Level: N/A
    Comments: Low moisture castor oil
Glycerine 99.5%:
    Functionality: Polyol   Use Level: N/A
    Comments: Vegetable and tallow grades available.
Jenkinol L 271:
    Functionality: External   Use Level: 0.3-0.8 phr
    Comments: Release agent in ejection molded polyurethane.

Application: Color Concentrates:
Jenkinol L 215:
    Functionality: Dispersant   Use Level: 2-5%
    Comments: Dispersant for various inorganic pigments.
Jenkinol L 221:
    Functionality: Dispersant   Use Level: 2-5%
    Comments: Dispersant for various inorganic pigments.

*FDA Approved

Axel Plastics Research Laboratories, Inc.: MOLDWIZ Pultrusion
Internal Lubricants:

INT-PS125:
   Active Ingredients: 100%
   Specific Gravity @ 25C: 0.95
   Viscosity @ 25C: 190 cps
   pH: 1.5-2.5
INT-PUL124:
   Active Ingredients: 100%
   Specific Gravity @ 25C: 0.96
   Viscosity @ 25C: 180 cps
   pH: 4.0-4.5
INT-EQ6:
   Active Ingredients: 100%
   Specific Gravity @ 25C: 0.925
   Viscosity @ 25C: 130 cps
   pH: 6.0-7.0
INT-ER8:
   Active Ingredients: 100%
   Specific Gravity @ 25C: 0.98
   Viscosity @ 25C: 515 cps
   pH: 6.0-7.0
INT-54:
   Active Ingredients: 100%
   Specific Gravity @ 25C: 0.90
   Viscosity @ 25C: 800 cps
   pH: 7.0-9.0
INT-1846:
   Active Ingredients: 100%
   Specific Gravity @ 25C: 0.986
   Viscosity @ 25C: 400 cps
   pH: 1.0-3.0
INT-1846N:
   Active Ingredients: 100%
   Specific Gravity @ 25C: 1.05
   Viscosity @ 25C: 800 cps
   pH: 6.0-8.0
INT-1866:
   Active Ingredients: 100%
   Specific Gravity @ 25C: 0.95
   Viscosity @ 25C: 130 cps
   pH: 8.0-10.0
INT-1850HT:
   Active Ingredients: 100%
   Specific Gravity @ 25C: 1.02
   Viscosity @ 25C: 300 cps
   pH: 1.5-2.5
INT-1890M:
   Active Ingredients: 100%
   Specific Gravity @ 25C: 0.994
   Viscosity @ 25C: 450 cps
   pH: 6.0-7.0

Chemax, Inc.: CHEMSTAT Lubricants:

327:
    Use Level % by Weight: PVC: 0.50-1.0//PS: 0.50-1.0
    Chemical nature: Ethylene bisstearamide
    Form: Beads
    FDA: Yes

G-118/42:*
    Use Level % by Weight: PE: 0.05-0.30//PP: 0.20-0.60
        PS: 0.20-0.60//Nylon: 2.0-5.0//SAN: 1.0-2.0//PU: 2.0-5.0
    Chemical Nature: GMS 42% Minimum
    Form: Beads
    FDA: Yes

G-118/52:*
    Use Level % by Weight: PE: 0.05-0.30//PP: 0.20-0.60
        PS: 1.5-4.0//Nylon: 2.0-5.0//SAN: 1.0-2.0//PU: 2.0-5.0
    Chemical Nature: GMS 52% Minimum
    Form: Flakes
    FDA: Yes

G-118/95:
    Use Level % by Weight: PE: 0.05-0.30//PP: 0.20-0.60
        PS: 1.5-4.0//Nylon: 0.50-1.5//SAN: 1.0-2.0//PU: 2.0-5.0
    Chemical Nature: GMS 95% Minimum
    Form: Beads
    FDA: Yes

G-118/9501:
    Use Level % by Weight: PE: 0.05-0.30/PP: 0.20-0.60
        PS: 1.5-4.0//Nylon: 0.50-1.5//SAN: 1.0-2.0//PU: 2.0-5.0
    Chemical Nature: GMS 95% Minimum
    Form: Powder
    FDA: Yes

G-1500:
    Use Level % by Weight: PE: 0.05-0.30//PP: 0.20-0.60
        Nylon: 2.0-5.0//SAN: 1.0-2.0//PU: 2.0-5.0
    Chemical Nature: Glycerol ester
    Form: Powder
    FDA: Yes

G-118/GTS:
    Use Level % by Weight: PS: 0.30-0.80
    Chemical Nature: Glycerol tristearate
    Form: Powder
    FDA: Yes

    *Available in kosher grade

Lonza Inc.: GLYCOLUBE Lubricants:

Glycolube 100:
    Chemical Description: Polyol Ester
    Internal lubricant for thermoplastics

Glycolube 110:
    Chemical Description: Polyol Ester
    Internal lubricant and antistat for PP and PE

Glycolube 110D:
    Food Grade
    Internal lubricant for rigid and semi-rigid filled PVC compounds.

Glycolube 140:
    Proprietary lubricant, anti-static agent
    Used primarily as an anti-static agent for polyethylene
and polypropylene compounds.
    A highly efficient internal lubricant for PVC

Glycolube 140 Kosher:
    Proprietary lubricant, anti-static agent
    Used primarily as an anti-static agent for polyethylene
and polypropylene compounds
    A highly efficient internal lubricant for PVC.

Glycolube 180:
    Polyol Ester
    Internal lubricant for PVC sheet and film, antistat for PP
and PE

Glycolube 674:
    Polyol Ester
    Internal lubricant recommended for rigid PVC, ABS, and
polystyrenic compounds

Glycolube 740 VEG FG:
    Proprietary Lubricant
    A highly efficient internal lubricant/co-stabilizer for
clear PVC compounds.
    May also be used as an anti-fog agent in plasticized PVC
formulations
    FDA approved for food contact use

Glycolube 742:
    Polyol Ester
    Internal and external lubricant for PVC.

Lonza Inc.: GLYCOLUBE Lubricants (Continued):

Glycolube 825:
   Polyol Ester
   Internal lubricant for rigid PVC, antistat for PP, PE

Glycolube 825 Kosher:
   Proprietary Lubricant, Anti-Static Agent
   An internal lubricant for rigid PVC compounds, providing a
good balance of thermal stability, melt flow and optical properties in a variety of systems.

Glycolube 853:
   Proprietary Lubricant
   FDA approved for food contact use

Glycolube P:
   Proprietary Lubricant
   Effective lubricant for many polymer compounds in both
extrusion and molding processes.

Glycolube P (ETS):
   Polyol Ester
   External lubricant for polycarbonate.

Glycolube PG:
   Plastics Additive--Food Grade
   External lubricant for a variety of PVC applications

Glycolube SG-1:
   Proprietary Lubricant
   Developed to meet the needs of rigid PVC compounds using
twin screw extrusion equipment.

Glycolube TS:
   Synthetic Wax
   External lubricant for engineering thermoplastics

Glycolube VL:
   Synthetic Wax
   Internal and external lubricant for thermoplastics

VINYLUBE 38:
   Ester Blend
   Lubricant/Antistat

Morton Plastics Additives: LUBRIOL Lubricants:

BF 4:
   Action: Internal
   Appearance: Solid
   Melting Range C: 74-80
   Compatible lubricant for clear rigid (blow and injection moulding, extrusion and calendering)
BF 10:
   Action: Internal
   Appearance: Solid
   Melting Range C: 62-70
   Highly compatible lube to be preferably used in stretch and conventional blow moulding
44 OL:
   Action: Internal
   Appearance: Liquid
   Density (25C): 0.950
   Internal allround lubricant, mainly suitable for soft applications
61:
   Action: Internal
   Appearance: Liquid
   Density (25C): 0.930
   Internal lubricant with medium compatibility; standard for rigid calendering
319:
   Action: Internal
   Appearance: Solid
   Melting Range C: 47-53
   Internal for rigid PVC: film extrusion, blow and injection moulding (both Sn and Pb stabilized)
77:
   Action: Internal/External
   Appearance: Liquid
   Density (25C): 0.860
   Balanced lubricant mainly for glass clear sheeting and plasticized items
200:
   Action: Internal
   Appearance: Solid
   Melting Range C: 42-47
   Almost universal internal lubricant for rigid PVC; can be utilized with different stabilizing systems
300S:
   Action: Internal/External
   Appearance: Solid
   Melting Range C: 53-59
   Balanced lubricant for rigid and semirigid extrusion & injection moulding

Morton Plastics Additives: LUBRIOL Lubricants (Continued):

E128OH:
    Action: External
    Appearance: Liquid
    Density (25C): 0.970
    Effective release agent to be preferably used for rigid and
plasticized calendering
E129:
    Action: External
    Appearance: Solid
    Melting Range C: 52-64
    Highly effective release agent specifically for rigid calen-
dered films
E190:
    Action: External
    Appearance: Solid
    Melting Range C: 77-83
    External lubricant with good release properties for rigid
calendering and blow moulding
E230:
    Action: External
    Appearance: Solid
    Melting Range C: 48-52
    External lube with release for PVC rigid calendering and
CPVC injection
E270:
    Action: Internal/External
    Appearance: Solid
    Melting Range C: 42-45
    Compatible with release for rigid PVC; blow moulding sheeting
and film
P163:
    Action: External
    Appearance: Solid
    Melting Range C: 107-117
    Release agent containing metal soaps suitable for CaZn blow
moulding and "Calanderette" film
P182:
    Action: External
    Appearance: Solid
    Melting Range C: 109-113
    Excellent release, particularly suitable for Sn rigid calen-
dering
SW/PF:
    Action: Internal/External
    Appearance: Solid
    Melting Range C: 140-142
    Balanced lubricant, effective organic antiblocking for rigid,
semirigid and plasticized PVC

Morton Plastics Additives: Blends:

Lubriol CL:
   Action: Internal/External
   Appearance: Liquid
   Food Appr.: BGA
   Balanced combinations, mainly customer tailored for rigid
and semirigid PVC

Lubriol CS:
   Action: Internal/External
   Appearance: Solid
   Food Appr.: BGA, FDA
   Solid combinations; same applications as Lubriol CL

Lubristat ST:
   Appearance: Liquid
   Food Appr.: BGA
   Blends of Sn stabilizers and lubricants (ev. metal soaps)
for rigid calendering

Lubristab ZC O:
   Appearance: Solid
   Food Appr.: BGA, FDA
   Range of one-packs containing CaZn stabilizers and lubricants
for calendered food grade films.

Lubristab ZC B:
   Appearance: Solid
   Food Appr.: BGA, FDA
   Range of one-packs containing CaZn stabilizers and proper
lubricants for food grade bottles

Penreco: PENRECO Petrolatums: For Plastics & Elastomers:

White Petrolatum USP:
Penreco Snow:
   Melting Point, F: 125/135
   Viscosity SUS @ 210F: 64/75
   Consistency @ 77F: 170/205
   Typical Congealing Point F: 123

Petrolatum USP:
Penreco Amber:
   Melting Point, F: 125/135
   Viscosity SUS @ 210F: 68/82
   Consistency @ 77F: 175/205
   Typical Congealing Point F: 123

Technical Petrolatum:
Penreco Red:
   Melting Point, F: 120/135
   Viscosity SUS @ 210F: 70/82
   Color: Red
   Consistency @ 77F: 175/205

Penreco 1520:
   Melting Point, F: 115/135
   Viscosity SUS @ 210F: 70/115
   Color: Dark Brown
   Consistency @ 77F: 170/260

Penreco 3070:
   Melting Point, F: 125/140
   Viscosity SUS @ 210F: 70/95
   Color: Dark Brown
   Consistency @ 77F: 130/175

Penreco 1180:
   Melting Point, F: 138/148
   Consistency @ 77F: 60/120

Petrolatum Applications:
Plastics & Elastomers:
   PVC-External Lubricants
   Rubber Processing Aids

Penreco: White Mineral Oils: For Plastics & Elastomers:

Mineral Oil USP:
Drakeol Supreme:
   Viscosity: SUS @ 100F: 520/570
   API @ 60F: 28.5/31.8
   Specific Gravity @ 77F: 0.860/0.878
   Flash Point: F/C: 490/254
   Pour Point: F/C: 20/-6
Drakeol 350:
   Viscosity: SUS @ 100F: 350/370
   API @ 60F: 29.8/32.8
   Specific Gravity @ 77F: 0.857/0.873
   Flash Point: F/C: 400/204
   Pour Point: F/C: 15/-9
Drakeol 35:
   Viscosity: SUS @ 100F: 340/365
   API @ 60F: 28.0/31.1
   Specific Gravity @ 77F: 0.864/0.881
   Flash Point: F/C: 435/224
   Pour Point: F/C: 5/-15
Drakeol 34:
   Viscosity: SUS @ 100F: 370/410
   API @ 60F: 29.7/32.3
   Specific Gravity @ 77F: 0.858/0.872
   Flash Point: F/C: 475/246
   Pour Point: F/C: 15/-9
Drakeol 32:
   Viscosity: SUS @ 100F: 312/330
   API @ 60F: 28.9/32.5
   Specific Gravity @ 77F: 0.856/0.876
   Flash Point: F/C: 415/213
   Pour Point: F/C: 10/-12
Drakeol 21:
   Viscosity: SUS @ 100F: 200/215
   API @ 60F: 28.9/33.2
   Specific Gravity @ 77F: 0.853/0.876
   Flash Point: F/C: 380/193
   Pour Point: F/C: 10/-12
Drakeol 19:
   Viscosity: SUS @ 100F: 180/190
   API @ 60F: 28.9/33.4
   Specific Gravity @ 77F: 0.852/0.876
   Flash Point: F/C: 370/188
   Pour Point: F/C: 10/-12

Plastics & Elastomers:
   Polystyrene-Internal Lubricants/PVC-External Lubricants
   Plastics Annealing/Catalyst Carriers
   Thermoplastic Rubber-Extender Oils

Witco Corp.: Lubricants for PVC:

Witco offers a wide variety of PVC lubricants including metallic stearates, amide waxes, and fatty acids. Witco supplies these products to the industry worldwide.

Calcium Stearates:
Witco calcium stearates are primarily used as internal lubricants and co-stabilizers in rigid PVC. But their unique solubility enables them to impart a degree of external lubrication as well. Their molecular polarity, particle configuration, and size distribution are qualities necessary for easy dispersion.

They have demonstrated excellent performance characteristics in rigid PVC compounding, extruding, and molding, where they serve to reduce friction and plate-out, improve powder-blend flow, regulate fusion rates, and enhance heat stability.

Witco calcium stearates can also reduce the possibility of "bloom," and the resultant color distortion, because they resist migration to the surface of finished products during storage.

Composites:
Witco also offers lubricant composites, which allow more efficient and economical handling of the components. These composites of stearates with either paraffin wax or ethylenebisstearamide are designed for use in extrusion or injection molding of rigid PVC.

Stearic Acid:
Hystrene 5016 stearic acid, triple pressed and characterized by its light color, is the finest available. At the correct concentratiion (typically 0.25-0.50 phr) it is the most effective lubricant for plasticized vinyl compounds stabilized with mixed-metal liquid stabilizers. In addition to supplying lubrication, Hystrene 5016 acts synergistically to improve the initial color and long-term stability of the vinyl compound. It also greatly reduces the amount of plate-out experienced with some liquid metallic stabilizers.

Amides:
Kemamide W-40 ethylenebisstearamide, a process lubricant for rigid PVC, is used in extruded house siding compounds and injection molded products such as pipe fittings. It contributes to the processing of rigid profiles by reducing the tendency of "screw sticking," minimizing surging, and increasing output rates. Typical use-levels are 1 to 2 phr.

Kemamide E erucamide is used in vinyl plastisol cap liners to provide the "release" properties that facilitate removing the caps from bottles.

Section XIII
Micas

Franklin Industrial Minerals: Mica and its Function in Plastic Composites:

The word "mica" is a generic term used to describe a group of complex hydrous potassium aluminum silicate minerals. They differ in chemical composition but share a unique laminar crystalline structure. Mica develops in a book-like form. Individual platelets have perfect basal cleavage which permits delaminating into extremely thin, high aspect ratio particles. These particles are tough and flexible.

Franklin Industrial Minerals mines and processes muscovite mica in Kings Mountain, NC, and in Velarde, NM. Products are available in particle size ranges from 15 to 300 microns.

Major Plastics Markets:
*Automotive: Large volumes of mica are currently used in production of blow molded seat backs for cars and trucks. Injection molded parts include glove boxes, battery trays, fan shrouds, fender liners, and instrument cluster housings. Mica-reinforced polyurethane/polyurea has been used success-fully since 1994 in automotive fascia and fenders.
Long term heat aging (LTHA) is important for under-the-hood applications. Detroit requires that parts function after 400 hours at 149 degrees C (300 degrees F). Mica enables plastic composites to retain their shape at these high temperatures.
*Construction and Appliance Markets: Mica reinforced compos-ites are used to replace metal and reduce the weight and cost of construction materials and appliances. Composites containing mica can be designed for superior stiffness and excellent dimensional stability.

Mica Benefits:
Composite Property: Flexural Modulus
Change Produced by Mica: Increased more than any other mineral
Composite Property: HDT
Change Produced by Mica: Increased more than any other mineral
Composite Property: Tensile and Flexural Strength
Change Produced by Mica: Greatly increased if a maleic anhydride modified PP additive is used
Composite Property: Warpage
Change Produced by Mica: Essentially eliminated
Composite Property: CLTE
Change Produced by Mica: Reduced substantially
Composite Property: Shrinkage
Change Produced by Mica: Greatly reduced from pure resin values
Composite Property: Chemical Resistance
Change Produced by Mica: Very high
Composite Property: Creep
Change Produced by Mica: Decreases significantly
Composite Property: Permeability
Change Produced by Mica: Greatly reduced
Composite Property: Surface Finish
Change Produced by Mica: Correct grades give class "A" surface

Franklin Industrial Minerals: Muscovite Mica Products:

Kings Mountain, NC Operation:
Wet Ground Mica Products:
WG-325:
    Median Size Microns: 33
    Aspect Ratio: 10
    BET Surface Area m2/gm: 4.7
    Bulk Density lbs/ft3: 12

HIMOD-270:
    Median Size Microns: 35
    Aspect Ratio: 15
    BET Surface Area m2/gm: 4.3
    Bulk Density lbs/ft3: 9

HAR-160:
    Median Size Microns: 39
    Aspect Ratio: 13
    BET Surface Area m2/gm: 4.9
    Bulk Density lbs/ft3: 9

WG-160:
    Median Size Microns: 39
    Aspect Ratio: 11
    BET Surface Area m2/gm: 4.7
    Bulk Density lbs/ft3: 12

HiMod-360:
    Median Size Microns: 41
    Aspect Ratio: 14
    BET Surface Area m2/gm: 5.8
    Bulk Density lbs/ft3: 9

Dry Ground Mica Products:
4-K:
    Median Size Microns: 50
    Aspect Ratio: 8
    BET Surface Area m2/gm: 3.6
    Bulk Density lbs/ft3: 16

1-K:
    Median Size Microns: 55
    Aspect Ratio: 9
    BET Surface Area m2/gm: 2.9
    Bulk Density lbs/ft3: 17

100-K:
    Median Size Microns: 81
    Aspect Ratio: 8
    BET Surface Area m2/gm: 2.5
    Bulk Density lbs/ft3: 18

Flake Mica Products:
L-140:
    Median Size Microns: 60
    Aspect Ratio: 13
    BET Surface Area m2/gm: 2.7
    Bulk Density lbs/ft3: 17

L-135:
    Median Size Microns: 230
    Aspect Ratio: 16
    BET Surface Area m2/gm: 1.2
    Bulk Density lbs/ft3: 13

20-K:
    Median Size Microns: 235
    Aspect Ratio: 9
    BET Surface Area m2/gm: 2.1
    Bulk Density lbs/ft3: 37

L-125:
    Median Size Microns: 460
    Aspect Ratio: 16
    BET Surface Area m2/gm: 1.4
    Bulk Density lbs/ft3: 22

Franklin Industrial Minerals: Muscovite Mica Products
  (Continued):

Kings Mountain, NC Operation (Continued):
Micro Mica Products:
C-4000:
    Median Size Microns: 17
    Aspect Ratio: 6
    BET Surface Area m2/gm: 6.3
    Bulk Density lbs/ft3: 12

C-1000:
    Median Size Microns: 26
    Aspect Ratio: 9
    BET Surface Area m2/gm: 5.3
    Bulk Density lbs/ft3: 12

C-3000:
    Median Size Microns: 23
    Aspect Ratio: 8
    BET Surface Area m2/gm: 5.4
    Bulk Density lbs/ft3: 12

C-500:
    Median Size Microns: 30
    Aspect Ratio: 9
    BET Surface Area m2/gm: 4.6
    Bulk Density lbs/ft3: 12

Valarde, NM Operation:
Dry Ground Mica Products:
PM-325:
    Median Size Microns: 24
    Aspect Ratio: 5
    BET Surface Area m2/gm: 3.9
    Bulk Density lbs/ft3: 17

MW-200:
    Median Size Microns: 48
    Aspect Ratio: 5
    BET Surface Area m2/gm: 1.9
    Bulk Density lbs/ft3: 21

MW-260:
    Median Size Microns: 37
    Aspect Ratio: 6
    BET Surface Area m2/gm: 2.5
    Bulk Density lbs/ft3: 19

Flake Mica Products:
MW-1117:
    Median Size Microns: 230
    Aspect Ratio: 5
    BET Surface Area m2/gm: 0.6
    Bulk Density lbs/ft3: 50

V-115:
    Median Size Microns: 550
    Aspect Ratio: 14
    BET Surface Area m2/gm: 0.4
    Bulk Density lbs/ft3: 51

Franklin Industrial Minerals: Muscovite Mica for Plastics
Applications:

Muscovite mica improves flexural modulus, heat distortion resistance and dimensional stability of thermoplastic and thermoset composites. Applications are established in polyolefin, polyamide, polyester and polyurethane/polyurea polymers.

Mica Product:

L-135:
  Aspect Ratio: 18
  Median Size Microns: 230
  Loose Bulk Density lb/ft3: 14

HiMod-270:
  Aspect Ratio: 15
  Median Size Microns: 35
  Loose Bulk Density lb/ft3: 10

HiMod-360:
  Aspect Ratio: 14
  Median Size Microns: 39
  Loose Bulk Density lb/ft3: 9

L-140:
  Aspect Ratio: 13
  Median Size Microns: 60
  Loose Bulk Density lb/ft3: 11

WG-325:
  Aspect Ratio: 10
  Median Size Microns: 33
  Loose Bulk Density lb/ft3: 12

4K:
  Aspect Ratio: 8
  Median Size Microns: 54
  Loose Bulk Density lb/ft3: 16

C-4000:
  Aspect Ratio: 6
  Median Size Microns: 17
  Loose Bulk Density lb/ft3: 12

PM-325:
  Aspect Ratio: 5
  Median Size Microns: 24
  Loose Bulk Density lb/ft3: 17

MW-200:
  Aspect Ratio: 5
  Median Size Microns: 48
  Loose Bulk Density lb/ft3: 21

Harwick Standard Distribution Corp.: Dry Ground Mica:

Typical Screen Analysis:

160-D:
    On 100 Mesh, %: Trace
    on 200 Mesh, %: 1-5
    On 325 Mesh, %: 18-26
    Pan, %: 74-82
    Density, lbs/cu ft: 12-16

325-MF:
    On 100 Mesh, %: 0.5
    On 200 Mesh, %: 6.0
    On 325 Mesh, %: 18-26
    Pan, %: 74-82
    Density, lbs/cu ft: 12-16

325-D:
    On 100 Mesh, %: Trace
    On 200 Mesh, %: 1.0
    on 325 Mesh, %: 11-18
    Pan, %: 82-89
    Density, lbs/cu ft: 12-16

325-FF:
    On 100 Mesh, %: Trace
    On 200 Mesh, %: 1.0
    On 325 Mesh, %: 4-10
    Pan, %: 90-96
    Density, lbs/cu ft: 12-16

325 Mesh:
    On 100 Mesh, %: 0.5
    On 200 Mesh, %: 6.0
    On 325 Mesh, %: 11-18
    Pan, %: 82-89
    Density, lbs/cu ft: 12-16

Typical Composition:
    These materials are ground flakes, silver-gray in color.
Their composition is Potassium Aluminum Silicate having the
following typical chemical composition and analysis:
Dry Ground Mica for Rubber:
    Silica (SiO_2), %: 45.1
    Alumina (Al_2O_3), %: 36.6
    Potash (K_2O), %: 11.8
    Magnesia (MgO), %: ----
    Iron Oxide (Fe_2O_3), %: Trace to 3.0
    Phosphorous (P), %: Trace
    Loss on Ignition, % Maximum: 5.0
    Moisture, % Maximum: 0.5

Intercorp Inc.: Mica-Wet Ground-KEMOLIT GS (Partial grade Listing):

Grades:
60:
    Typical Aspect Ratio: 40:1
    Typical Sieve Analysis: % Passing-325 mesh: 38%
    Grade Note: Coarse, High Aspect Ratio (HAR)

180:
    Typical Aspect Ratio: 40:1
    Typical Sieve Analysis, % Passing-325 mesh: 64%
    Grade Note: HAR

325:
    Typical Aspect Ratio: 40:1
    Typical Sieve Analysis, % Passing-325 mesh: 93%
    Grade Note: HAR

3X:
    Typical Aspect Ratio: 40:1
    Typical Sieve Analysis, % Passing-500 mesh: 98%
    Grade Note: HAR, particle suspension aid

Applications:
    Thermosets & Thermoplastics Compounds, Paints & Coatings, Insulation

Mica-Surface Modified Wet Ground-FILLEX SD (Partial Grade Listing):

160:
    Typical Aspect Ratio: 40:1
    Typical Sieve Analysis: % Passing 200 Mesh: 55%
    Grade Note: High aspect ratio

325:
    Typical Aspect Ratio: 40:1
    Typical Sieve Analysis: % Passing 200 Mesh: 95%
    Grade Note: High aspect ratio, highest DOI

Applications:
    Eng. resins incl. olefins, nylons, polyesters & thermosets-urethane, phenolic, polyester, etc.; RRIM & injection molding.

Pacer Technology, Inc.: MICAFLEX R200:

Typical Sieve Analysis:
```
    On U.S.S. #80:      0.1%
       U.S.S. #100:     0.5%
       U.S.S. #140:    10.0%
       U.S.S. #200:    20.0%
       U.S.S. #325:    33.0%
    Below U.S.S #325 in Pan: 36.4%
```

Typical Physical/Chemical Properties:
    Bulk Density: 35-45 lbs./cu.ft.
    Loss on Ignition: 1.5-2.5
    Loss on Drying: 0.01
    Specific Gravity: 2.75
    Refractive Index: 1.6
    pH: 8.0
    Moh's Hardness: 2.5
    Free Respirable Silica: 0.261%
    Stable, virtually inert except to hydrofluoric and concen-
trated sulfuric acid.
    Impervious to water and atmosphere.
    Unaffected by exposure to extended low and high temperatures.

Typical Chemical Analysis:
```
    Silica (SiO2)         53.4%
    Alumina (Al2O3)       26.1%
    Iron (Fe2O3)           2.3%
    Iron (FeO)             2.7%
    Soda (Na2O)            2.1%
    Potash (K2O)           7.6%
    Calcium (CaO)          0.3%
    Magnesium (MgO)        1.4%
```

Section XIV
Pigments, Colorants, and Dyes

Akzo Nobel: KETJENBLACK Conductive Carbon Blacks:

Ketjenblack EC-300J Conductive Carbon Black:
CAS Reg. No.: 1333-86-4

Ketjenblack EC-300J is a carbon black having a unique morphology. It is extremely suitable for electroconductive applications. Due to its structure only one third the amount of Ketjenblack EC-300J is needed compared to conventional carbon blacks in many applications. Ketjenblack EC-300J is especially useful in applications requiring low ash and is recommended for use in material which must have a smooth surface.

Product Properties:
Iodine absorption, mg/g: 740-840
Pore volume (DBP-absorption), cm3/100g: 310-345
Ash content, max. %: 0.1
Grit Content, mg/kg max.: 30
Volatiles, max. %: 1.0
Moisture, max. %: 0.5
Apparent bulk density, kg/m3: 125-145
pH: 8-10

Ketjenblack EC-600 JD Conductive Carbon Black:
CAS Reg. No.: 1333-86-4

Ketjenblack EC-600 JD is a carbon black having a unique morphology. It is extremely suitable for electroconductive applications. Due to its structure only one sixth the amount of Ketjenblack EC-600 JD is needed compared to conventional carbon blacks in many applications. Ketjenblack EC-600 JD is especially useful in applications requiring extremely high conductivity at relatively low loadings of carbon black.

Product Properties:
Iodine absorption, mg/g: 1000-1150
Pore volume (DBP-absorption), cm3/100g: 480-510
Ash content, max. %: 0.1
Grit content, mg/kg max.: 30
Volatiles, max. %: 1.0
Moisture, max. %: 0.5
Apparent bulk density, kg/m3: 100-120
pH: 8-10

BASF Corp.: Dyes for Plastics: SUDAN/LUMOGEN/THERMOPLAST:

Sudan Dyes:
Sudan Yellow 146:
    Type of Dye: Azo dye
    Color Index: Solvent Yellow 16/12700
    Heat Stability (GPPS): 280C
    Fastness to light (1/3 SD GPPS): 8 (8=best fastness to light)
    Specific Gravity: 1.29 cm3

RPVC: +	PS: +	PA: 0
SAN: 0	PMMA: +	PETP: -
PC: +	SB: -	CA/CAB: -
ABS: -	ASA: -	UP: -
MF/PF: -		

    0 = Suitable under certain conditions only, preliminary trials
        recommended
    + = Suitable   - = Unsuitable

Sudan M Yellow 150*:
    Type of Dye: Azo dye
    Color Index: Solvent Yellow 56/11021
    Heat Stability: (GPPS): 240C
    Fastness to Light (1/3 SD GPPS): 2 (8=best fastness to light)
    Specific Gravity: 1.15 gm3

PVC: +	PS: +	PA: -
SAN: 0	PMMA: -	PETP: -
PC: -	SB: -	CA/CAB: -
ABS: -	ASA: -	UP: -
MF/PF: -		

    0 = Suitable under certain conditions only, preliminary trials
        recommended
    + = Suitable   - = Unsuitable
    *listed on the German ban of Azo dyes

Sudan M Red 380*:
    Type of Dye: Azo dye
    Color Index: Solvent Red/26105
    Heat Stability (GPPS): 260C
    Fastness to Light (1/3 SD GPPS): 7 (8=best fastness to light)
    Specific Gravity: 1.29 gm3

PVC: +	PS: +	PA: -
SAN: 0	PMMA: 0	PETP: -
PC: 0	SB: -	CA/CAB: -
ABS: -	ASA: -	UP: -
MF/PF: -		

    0 = Suitable under certain conditions only, preliminary trials
        recommended
    + = Suitable   - = Unsuitable
    * Listed on the German ban of Azo dyes

BASF Corp.: Dyes for Plastics: SUDAN/LUMOGEN/THERMOPLAST (Continued):

Sudan Blue 670:
    Type of Dye: Azo dye
    Color Index: Solvent Blue/61554
    Heat Stability (GPPS): 280C
    Fastness to Light (1/3 SD GPPS): 7 (8=best fastness to light)
    Specific Gravity: 1.23 gm3

PVC: +	PS: +	PA: 0
SAN: 0	PMMA: +	PETP: -
PC: +	SB: +	CA/CAB: -
ABS: +	ASA: 0	UP: -

    MF/PF: -
    0 = Suitable under certain conditions only, preliminary trials
        recommended
    + = Suitable   - = Unsuitable

Lumogen F Yellow 083:
    Type of Dye: Perylene
    Color Index: -------
    Heat Stability (GPPS): 280C
    Fastness to weathering (PMMA): >90
    Specific Gravity: 1.27 cm3
    Max (nm) Absorption in ethylene dichloride: 476
    Max (nm) Absorption in PMMA: 473
    Fluorescence in ethylene dichloride: 490
    Max (nm) Quantum Yield: 0.91

Lumogen F Orange 240:
    Type of Dye: Perylene
    Color Index: --------
    Fastness to Weathering (PMMA): >85
    Heat Stability (PC): 300C
    Melting Point: >300
    Specific Gravity: 1.36 cm3
    Max (nm) Absorption in ethylene dichloride: 524
    Max (nm) Absorption in PMMA: 525
    Fluorescence (nm) in ethylene dichloride: 539
    Max Quantum Yield: 0.99

BASF Corp.: Dyes for Plastics: SUDAN/LUMOGEN/THERMOPLAST (Continued):

Lumogen F Red 300:
    Type of Dye: Perylene
    Color Index: --------
    Fastness to Weathering (PMMA): >95
    Heat Stability (PC): 300C
    Melting Point: >300C
    Specific Gravity: 1.40 cm3
    Max (nm) Absorption in ethylene dichloride: 578
    Max (nm) Absorption in PMMA: 578
    Fluorescence (nm) in ethylene dichloride: 613
    Max. Quantum Yield: 0.98
Lumogen F Violet 570:
    Type of Dye: Naphtalamide
    Color Index: Fluorescence Brightener 331
    Fastness to Weathering (PMMA): >80 after 50 days
    Heat Stability (PC): 300C
    Melting Point: >300
    Specific Gravity: 1.28 cm3
    Max (nm) Absorption in ethylene dichloride: 378
    Max (nm) Absorption in PMMA: 378
    Fluorescence (nm) in ethylene dichloride: 413
    Max. Quantum Yield: 0.94

Thermoplast Yellow 084:
    Type of Dye: Polycyclic
    Color Index: Solvent Green 5/59075
    Heat Stability (GPPS): 300C
    Fastness to Light (1/3 SD): 7 (8=best fastness to light)
    Specific Gravity: 1.27 cm3

RPVC: +	PS: +	PA: 0
SAN: +	PMMA: +	PETP: 0
PC: +	SB: +	CA/CAB: 0
ABS: +	ASA: +	UP: +

    MF/PF: +
    0= Suitable under certain conditions only, preliminary trials
      recommended
    += Suitable   - = Unsuitable

Thermoplast Yellow 104:
    Type of Dye: Pyrazolone Yellow
    Color Index: Solvent Yellow 93
    Heat Stability (GPPS): 300C
    Fastness to light (1/3 SD GPPS): 8 (8=best fastness to light)
    Specific Gravity: 1.31 gm3

PVC: +	PS: +	PA: 0
SAN: +	PMMA: +	PETP: 0
PC: +	SB: 0	CA/CAB: 0
ABS: 0	ASA: 0	UP: +

    MF/PF: +
    0= Suitable under certain conditions only, preliminary trials
      recommended
    += Suitable   - = Unsuitable

**BASF Corp.: Dyes for Plastics: SUDAN/LUMOGEN/THERMOPLAST
(Continued):**

Thermoplast Blue 684:
    Type of Dye: Anthraquinone Blue
    Color Index: Solvent Violet 13/60725
    Heat Stability (GPPS): 300C
    Fastness to light (1/3 SD GPPS): 7-8(8=best fastness to light)
    Specific Gravity: 1.38 gm3

PVC: +	PS: +	PA: 0
SAN: +	PMMA: +	PETP: 0
PC: +	SB: 0	CA/CAB: 0
ABS: 0	ASA: 0	UP: +

    MF/PF: +
    0 = suitable under certain conditions only, preliminary trials
        recommended
    + = Suitable   - = Unsuitable

Thermoplast Red LB 454:
    Type of Dye: Monoazo Red
    Color Index: Solvent Red 195
    Heat Stability (GPPS): 300C
    Fastness to Light (1/3 SD GPPS): 8 (8=best fastness to light)
    Specific Gravity: 1.28 gm3

PVC: +	PS: +	PA: 0
SAN: +	PMMA: +	PETP: 0
PC: +	SB: 0	CA/CAB: 0
ABS: 0	ASA: 0	UP: +

    MF/PF: +
    0 = suitable under certain conditions only, preliminary trials
        recommended
    + = Suitable   - = Unsuitable

Thermoplast Black X-70:
    Type of Dye: Mixture
    Color Index: -------
    Heat Stability (GPPS): 300C
    Fastness to Light (1/3 SD GPPS): 8 (8=best fastness to light)
    Fastness to Weathering (PS): 5
    Fastness to Migration (PS): +
    Specific Gravity: 1.36 gm3

PVC: +	PS: +	PA: 0
SAN: +	PMMA: +	PETP: 0
PC: +	SB: 0	CA/CAB: 0
ABS: 0	ASA: 0	UP: +

    MF/PF: +
    0 = Suitable under certain conditions only, preliminary trials
        recommended
    + = Suitable   - = Unsuitable

Cabot Corp.: CABOT Special Blacks for Plastics:

Fluffy/Pellets:
High Color:
MONARCH 1100/BLACK PEARLS 1100:
    Jetness Index: 65
    For coloring-gives highest jetness

Monarch 900/Black Pearls 900:
    Jetness Index: 70
    For coloring-gives high jetness

Medium Color:
Monarch 880/Black Pearls 880:
    Jetness Index: 74
    For coloring-med. jetness-exc. dispersion

Monarch 800/Black Pearls 800:
    Jetness Index: 73
    For coloring-med. jet-low oil absorption

Monarch 700/Black Pearls 700:
    Jetness Index: 78
    For coloring-med. jet-exc. dispersion

Aftertreated:
Monarch 1000/Black Pearls 1000:
    Jetness Index: 69
    For resist. plastics-high jetness

MOGUL L/Black Pearls L:
    Jetness Index: 83
    For resistive plastics

REGAL 400R/Regal 400:
    Jetness Index: 84
    For higher dielectric PVC cble cpd

Conductive:
VULCAN XC72R/VULCAN XC72:
    Jetness Index: 87
    For optimum electrical conductivity

-----/Vulcan PA80:
    Jetness Index: 88
    For UV protection and conductivity

Vulcan PF*/Vulcan P:
    Jetness Index: 88
    For good electrical conductivity

-----/ELFTEX TP*:
    Jetness Index: 88
    For good electr. conductivity and UV protection

Cabot Corp.: CABOT Special Blacks for Plastics (Continued):

Fluffy/Pellets:
Regular Color:
-----/Vulcan 9A32:
    Jetness Index: 84
    U.S industry standard for UV protection

Elftex 675*/Elftex 670*:
    Jetness Index: 87
    Very good UV protection

Regal 660R/Regal 660:
    Jetness Index: 83
    For color and high tint

-----/Elftex 570*
    Jetness Index: 88
    Good UV protection-good dispersion

-----/Black Pearls 570:
    Jetness Index: 86
    Good UV protection-good dispersion

Eltex 465*/Elftex 460*:
    Jetness Index: 90
    High tint strength-good dispersion

-----/Black Pearls 460:
    Jetness Index: 87
    High tint-blue tone

Regal 330R/Regal 330:
    Jetness Index: 84
    High tint-good dispersion & color

Regal 300R/Black Pearls 450:
    Jetness Index: 87
    High tint-good dispersion & color

Elftex 435*/Elftex 430*:
    Jetness Index: 87
    Low structure for MB or concentrates

Elftex 415*/-----:
    Jetness Index: 84
    High tint with blue tone

    *Grades produced only in Europe

Cabot Corp.: CABOT Special Blacks for Plastics (Continued):

Fluffy/Pellets:
Utility:
Elftex 285*/Elftex 280*:
    Jetness Index: 95
    Blue tone-exc. dispersion-surface smooth

-----/Black Pearls 280:
    Jetness Index: 95
    Blue tone-exc. dispersion-surface smooth

-----/Elftex 160*:
    Jetness Index: 96
    Very blue tone-easy dispersion

-----/Black Pearls 130:
    Jetness Index: 99
    Excellent jetness & tint at low cost

Elftex 125*/Elftex 120*:
    Jetness Index: 94
    Low price blk. for highly loaded concentrates

Monarch 120/Black Pearls 120:
    Jetness Index: 99
    For concentrates--low cost, good color, blue tint

*Grades produced only in Europe

Cabot Corp.: FDA-Compliant Carbon Black For Indirect Food Contact Applications:

BLACK PEARLS 4350 Carbon Black:

FDA regulation 21CFR 178.3297 Colorants for Polymers has been amended to allow the use of high-purity furnace blacks. Cabot's Black Pearls 4350 carbon black is now compliant with this amended regulation. This carbon black may be used at up to 2.5% loading to make articles intended for use in producing, manufacturing, packing, processing, preparing, treating, packaging, transporting or holding food under all temperatures.

Performance Features:

*Jetness: Provides a medium level of jetness and medium blue undertone similar to gas-fired channel blacks.

*Dispersion: Offers superior dispersion to gas-fired channel blacks.

*Impact Strength: Provides better polymer impact resistance than gas-fired channel black.

*Viscosity: Gives polymer viscosity comparable to that of channel-black-loaded polymer.

Black Pearls 4750 Carbon Black:

FDA Regulation 21CFR 178.3297 Colorants for Polymers has been amended to allow the use of high-purity furnace blacks. Black Pearls 4750 is now compliant with this amended regulation. This carbon black may be used at up to 2.5% loading to make articles intended for use in producing, manufacturing, packing, processing, preparing, treating, packaging, transporting or holding food under all temperatures.

Performance Features:

*Jetness: Exhibits higher jetness levels than gas-fired channel black.

*Dispersion: Offers superior dispersion to gas-fired channel blacks.

*Impact Strength: 4750 carbon-black-filled polymers exhibit improved impact strength vs. the channel black.

*Viscosity: Gives polymer viscosity comparable to that of channel-black-loaded polymer.

Clariant Corp.: GRAPHTOL Pigments for Plastics:
Graphtol pigments are somewhat inferior in general fastness properties to PV Fast Pigments.
The migration properties of the PV Fast pigments are not quite reached. The light fastness and heat stability meet the general requirements.

Yellow 3GP:
    C.I. Pigment Yellow 155
Yellow GG:
    C.I. Pigment Yellow 17
    C.I. No.: 21105
Yellow GXS:
    C.I. Pigment Yellow 14
    C.I. No.: 21095
Yellow GR:
    C.I. Pigment Yellow 13
    C.I. No.: 21100
Orange GPS:
    C.I. Pigment Orange 13
    C.I. No.: 21110
Orange RL:
    C.I. Pigment Orange 34
    C.I. No.: 21115
Red HFG:
    C.I. Pigment Orange 38
    C.I. No.: 12367
Red BB:
    C.I. Pigment Red 38
    C.I. No.: 21120
Red LG:
    C.I. Pigment Red 53:1
    C.I. No.: 15585:1
Red LC:
    C.I. Pigment Red 53:1
    C.I. No.: 15585:1
Fire Red 3RLP:
    C.I. Pigment Red 48:3
    C.I. No.: 15865:3
Red HF2B:
    C.I. Pigment Red 208
    C.I. No.: 12514
Red F3RK 70:
    C.I. Pigment Red 170
    C.I. No.: 12475
Red 2BN:
    C.I. Pigment Red 262
Red F5RK:
    C.I. Pigment Red 170
    C.I. No.: 12475
Carmine HF3C:
    C.I. Pigment Red 176
    C.I. No.: 12515
Carmine HF4C:
    C.I. Pigment Red 185
    C.I. No.: 12516

Rubine L6B:
    C.I. Pigment Red 57:1
    C.I. No.: 15850:1
Bordeaux HF3R:
    C.I. Pigment Violet 32
    C.I. No.: 12517

WATCHUNG: RT-430-D:
Watchung Red B:
    C.I. Pigment Red 48:2
    C.I. No.: 15865:2

Clariant Corp.: PV Fast Pigments for Plastics:

Because of their very good general fastness properties the
PV Fast pigments can, with few exceptions, be used in all types
of plastic. They were selected primarily on the strength of their
light fastness, heat stability, and migration fastness.

PV:
Fast Yellow H2G:
      C.I. Pigment Yellow 120
      C.I. No.: 11783
Fast Yellow HG:
      C.I. Pigment Yellow 180
      C.I. No.: 21290
Fast Yellow HGR:
      C.I. Pigment Yellow 191
      C.I. No.: 18795
Fast Yellow HR:
      C.I. Pigment Yellow 83
      C.I. No.: 21108
Fast Yellow HR 02:
      C.I. Pigment Yellow 83
      C.I. No.: 21108
Fast Yellow H3R:
      C.I. Pigment Yellow 181
      C.I. No.: 11777
Fast Orange H4GL 01:
      C.I. Pigment Orange 72
Fast Orange GRL:
      C.I. Pigment Orange 43
      C.I. No.: 71105
Fast Orange 6RL:
      C.I. Pigment Orange 68
Fast Scarlet 4RF:
      C.I. Pigment Red 242
      C.I. No.: 20067
Fast Red B:
      C.I. Pigment Red 149
      C.I. No.: 71137
Fast Red HFT:
      C.I. Pigment Red 175
      C.I. No.: 12513
Fast Red 3B:
      C.I. Pigment Red 144
      C.I. No.: 20735
Fast Red HG:
      C.I. Pigment Red 247
      C.I. No. 15915
Fast Red HB:
      C.I. PIgment Red 247
      C.I. No.: 15915

Clariant Corp.: PV Fast Pigments for Plastics (Continued):

Fast Red BNP:
  C.I. Pigment Red 214
Fast Red HF4B:
  C.I. Pigment Red 187
  C.I. No.: 12486
Fast Red E3B:
  C.I. Pigment Violet 19
  C.I. No.: 73900
Fast Red E5B:
  C.I. Pigment Violet 19
  C.I. No.: 73900
Fast Pink E:
  C.I. Pigment Red 122
  C.I. No.: 73915
Fast Pink E 01:
  C.I. Pigment Red 122
  C.I. No.: 73915
Fast Violet ER VP 2223:
  C.I. Pigment Violet 19
  C.I. No.: 73900
Fast Violet BLP:
  C.I. Pigment Violet 23
  C.I. No.: 51319
Fast Violet RL:
  C.I. Pigment Violet 23
  C.I. No.: 51319
Fast Blue A2R:
  C.I. PIgment Blue 15:1
  C.I. No.: 74160
Fast Blue BV:
  C.I. Pigment Blue 15:1
  C.I. No. 74160
Fast Blue 2GLSP:
  C.I. Pigment Blue 15:3
  C.I. No. 74160
Fast Blue BG:
  C.I. Pigment Blue 15:3
  C.I. No: 74160
Fast Green GG:
  C.I. Pigment Green 7
  C.I. No.: 74260
Fast Pigment Green GNX:
  C.I. Green 7
  C.I. No.: 74260
Fast Pigment Brown HFR:
  C.I. Brown 25
  C.I. No.: 12510
Fast White R 01:
  C.I. Pigment White 6
  C.I. No.: 77891

Day-Glo Color Corp.: DAY-GLO Pigments for Plastics:

A & AX Pigments:
Are thermoplastic fluorescent pigments which are recommended
for a wide range of applications where resistance to strong
solvents is not needed. The A- and AX-Pigments are used in such
applications as paper coatings, vinyl coated fabric, A-type
gravure inks, paints, screen inks, vinyl plastisols and organi-
sols and plastics with melt temperatures less than 380F (193C).
Available Colors:
Aurora Pink*
Neon Red*
Rocket Red*
Fire Orange*
Blaze Orange*
Arc Yellow*
Saturn Yellow*
Signal Green*
Horizon Blue*
Corona Magenta*
* Trademark of Day-Glo Color Corp.

T & GT Pigments:
Are based on a thermoset resin. They are useful in applica-
tions where solvent resistance is required. GT-Pigments have
higher color strength than T-Pigments.
Available Colors:
Aurora Pink*
Neon Red*
Rocket Red*
Fire Orange*
Blaze Orange*
Arc Yellow*
Saturn Yellow*
Signal Green*
Horizon Blue*
Corona Magenta*
* Trademark of Day-Glo Color Corp.

Day-Glo Color Corp.: DAZZLE COLORS:

A new line of highly visible and energetic polyester metallized foil films. Coated in a variety of vibrant colors, these
sparkle colors can be used in graphic arts, paint, textile
and plastics applications. These products are available in a
wide range of colors and particle sizes, tailored to specific
applications. The standard particle sizes are 0.4 mm and 0.1 mm
square cut for Dazzle Colors. Smaller or larger sizes are
available by special request.

Color:
General Purpose & Low
   Temperature:
      Silver
      Gold
      Green
      Violet
      Red
      Blue
      Pink
      Dark Blue
      Light Gold
      Cherry Pink

      Sky Blue
      Black
      White
      Multicolor
      Light Brown
      Dark Brown

Textile:
   Blizzard White
   Sunburst Yellow
   Forest Green
   Lavender Violet
   Cocktail Pink
   Aqua Blue
   Fire Engine Red

Aluminum:
   Anniversary Silver
   Treasure Gold
   Emerald Green
   Copper Orange
   Ruby Red
   Royal Blue
   Champagne Pink
   Midnight Black

Hologram:
   Jewel Hologram

Phosphorescent:
   Night Glow

Degussa Corp.: Carbon Blacks:

Domestic Furnace Blacks:
PRINTEX 60:
   Type: RCF: Regular Color Furnace
   Particle Size (nm): 21
   DBP Absorption (ml/100g): Powder: 116
                                    Beads: 100
   Relative Tinting Strength/IRB3=100: 105
   BET-Surface Area (m2/g): 115

Printex L:
   Type: Conductive Black
   Particle Size (nm): 23
   DBP Absorption (ml/100g): Powder: 116
                                    Beads: 114
   Relative Tinting Strength/IRB3=100: 102
   BET-Surface Area (m2/g): 150

Printex 55:
   Type: RCF
   Particle Size (nm): 25
   DBP Absorption (ml/100g): Powder: 48
                                    Beads: 48
   Relative Tinting Strength/IRB3=100: 119
   BET-Surface Area (m2/g): 110

Printex 45:
   Type: RCF
   Particle Size (nm): 26
   DBP Absorption (ml/100g): Powder: 52
                                    Beads: 52
   Relative Tinting Strength/IRB3=100: 115
   BET-Surface Area (m2/g): 90

Printex 300:
   Type: RCF
   Particle Size (nm): 27
   DBP Absorption (ml/100g): Powder: 65
                                    Beads: 68
   Relative Tinting Strength/IRB3=100: 109
   BET-Surface Area (m2/g): 80

Printex 30:
   Type: RCF
   Particle Size (nm): 27
   DBP Absorption (ml/100g): Powder: 106
                                    Beads: 99
   Relative Tinting Strength/IRB3=100: 100
   BET-Surface Area (m2/g): 80

Degussa Corp.: Carbon Blacks (Continued):

Domestic Furnace Blacks (Continued):
Printex 3:
   Type: RCF (Regular Color Furnace)
   Particle Size (nm): 27
   DBP Absorption (ml/100g): Powder: 125
                                    Beads: 124
   Relative Tinting Strength/IRB3=100: 90
   BET Surface Area (m2/g): 80

Printex 35:
   Type: RCF
   Particle Size (nm): 31
   DBP Absorption (ml/100g): Powder: 42
                                    Beads: 42
   Relative Tinting Strength/IRB3=100: 100
   BET Surface Area (m2/g): 65

Printex A:
   Type: LCF (Low Color Furnace)
   Particle Size (nm): 41
   DBP Absorption (ml/100g): Powder: 118
                                    Beads: 118
   Relative Tinting Strength/IRB3=100: 69
   BET Surface Area (m2/g): 45

Printex G:
   Type: LCF
   Particle Size (nm): 51
   DBP Absorption (ml/100g): Powder: 96
                                    Beads: 89
   Relative Tinting Strength/IRB3=100: 64
   BET Surface Area (m2/g: 30

Printex 25:
   Type: LCF
   Particle Size (nm): 56
   DBP Absorption (ml/100g): Powder: 46
                                    Beads: 46
   Relative Tinting Strength/IRB3=100: 88
   BET Surface Area (m2/g): 45

Degussa Corp.: Carbon Blacks (Continued):

Imported Furnace Blacks:
Printex 85:
    Type: MCF (Medium Color Furnace)
    Particle Size (nm): 16
    DBP Absorption (ml/100g): Powder: 48
                                   Beads: 46
    Relative Tinting Strength/IRB3=100: 120
    BET-Surface Area (m2/g): 200

Printex L6:
    Type: Conductive Black
    Particle Size (nm): 18
    DBP Absorption (ml/100g): Powder: 120
                                  Beads: 120
    Relative Tinting Strength/IRB=3: 108
    BET-Surface Area (m2/g): 265

Special Black 550:
    Type: RCF (Regular Color Furnace)
    Particle Size (nm): 25
    DBP Absorption (ml/100g): Powder: 49
    Relative Tinting Strength/IRB3=100: 114
    BET-Surface Area (m2/g): 110

Special Black 350:
    Type: RCF
    Particle Size (nm): 31
    DBP Absorption (ml/100g): Powder: 50
    Relative Tinting Strength/IRB3=100: 99
    BET-Surface Area (m2/g): 65

Special Black 100:
    Type: LCF (Low Color Furnace)
    Particle Size (nm): 50
    DBP Absorption (ml/100g): Powder: 93
    Relative Tinting Strength/IRB3=100: 64
    BET-Surface Area (m2/g): 30

Special Black 250:
    Type: LCF
    Particle Size (nm): 56
    DBP Absorption (ml/100g): Powder: 48
    Relative Tinting Strength/IRB3=100: 90
    BET-Surface Area (m2/g): 40

Printex XE 2:
    Type: Extra Conductive Black
    Particle Size (nm): 35
    DBP Absorption (ml/100g): Beads: 400
    Relative Tinting Strength/IRB3=100: 124
    BET-Surface Area (m2/g): 1000

Degussa Corp.: Carbon Blacks (Continued):

Channel Type Imported Blacks:
Color Black FW 200:
    Equivalent to: HCC (High Color Channel)
    Particle Size (nm): 13
    Oil Absorption (F.P.)%: Powder: 620
                        Beads: 450
    Relative Tinting Strength/IRB3=100: 107
    BET-Surface Area (m2/g): 460

Color Black FW 2:
    Equivalent to: HCC
    Particle Size (nm): 13
    Oil Absorption (F.P.)%: Powder: 670
                        Beads: 500
    Relative Tinting Strength/IRB3=100: 107
    BET-Surface Area (m2/g): 460

Color Black FW 18:
    Equivalent to: HCC
    Particle Size (nm): 15
    Oil Absorption (F.P.) %: Powder: 810
                        Beads: 600
    Relative Tinting Strength/IRB3=100: 122
    BET-Surface Area (m2/g): 260

Special Black 6:
    Equivalent to: HCC
    Particle Size (nm): 17
    Oil Absorption (F.P.) %: Powder: 520
                        Beads: 400
    Relative Tinting Strength/IRB3=100: 110
    BET-Surface Area (m2/g): 300

Color Black S 170:
    Equivalent to: MCC (Medium Color Channel)
    Particle Size (nm): 17
    Oil Absorption (F.P.) %: Powder: 680
                        Beads: 560
    Relative Tinting Strength/IRB3=100: 121
    BET-Surface Area (m2/g): 200

Special Black 5:
    Equivalent to: MCC
    Particle Size (nm): 20
    Oil Absorption (F.P.) %: Powder: 410
                        Beads: 360
    Relative Tinting Strength/IRB3=100: 110
    BET-Surface Area (m2/g): 240

Degussa Corp.: Carbon Blacks (Continued):

Channel Type Imported Blacks (Continued):
Special Black 4:
   Equivalent to: RCC (Regular Color Channel)
   Particle Size (nm): 25
   Oil Absorption (F.P.) %: Powder: 300
                                 Beads: 230
   Relative Tinting Strength/IRB3=100: 100
   BET Surface Area (m2/g): 180

Special Black 4 A:
   Particle Size (nm): 25
   Oil Absorption (F.P.) %: Powder: 230
   Relative Tinting Strength/IRB3=100: 95
   BET Surface Area (m2/g): 180

Printex U:
   Equivalent to: RCC
   Particle Size (nm): 25
   Oil Absorption (F.P.) %: Powder: 420
                                   Beads: 310
   Relative Tinting Strength/IRB3=100: 108
   BET Surface Area (m2/g): 100

Printex 150 T:
   Equivalent to: RCC
   Particle Size (nm): 29
   Oil Absorption (F.P.) %: Powder: 400
   Relative Tinting Strength/IRB3=100: 100
   BET Surface Area (m2/g): 110

Printex 140 U:
   Equivalent to: RCC
   Particle Size (nm): 29
   Oil Absorption (F.P.) %: Powder: 380
                                   Beads: 280
   Relative Tinting Strength/IRB3=100: 103
   BET Surface Area (m2/g): 90

Imported Lamp Blacks:
Lamp Black 101:
   Type: Lamp Black
   Particle Size (nm): 95
   DBP Absorption (ml/100g): Powder: 112
                                      Beads: 100
   Relative Tinting Strength/IRB3=100: 29
   BET-Surface Area (m2/g): 20

Degussa Corp.: Pigment Black Preparations for Plastics:

COLCOLOR E30/90:
    Pigment Black: Printex 90
    Pigment Black Concentration %: 30
    Binder: Polyethylene, low density
Colcolor E 40/P:
    Pigment Black: Printex P
    Pigment Black Concentration %: 40
    Binder: Polyethylene, low density
Colcolor E 40/60:
    Pigment Black: Printex 60
    Pigment Black Concentration %: 40
    Binder: Polyethylene, low density
Colcolor E 50/G:
    Pigment Black: Printex G
    Pigment Black Concentration %: 50
    Binder: Polyethylene, low density
Colcolor E 60/R:
    Pigment Black: LCF black
    Pigment Black Concentration %: 60
    Binder: Polyethylene, low density
Colcolor E-HD 25/L:
    Pigment Black: Printex L
    Pigment Black Concentration %: 25
    Binder: Polyethylene, high density
Colcolor LC PP 10:
    Pigment Black: Printex P
    Pigment Black Concentration %: 25
    Binder: Polypropylene
Colcolor P 30/18:
    Pigment Black: Colour Black FW 18
    Pigment Black Concentration %: 30
    Binder: Polypropylene
Colcolor EVA 40/60:
    Pigment Black: Printex 60
    Pigment Black Concentration %: 40
    Binder: Ethylene Vinyl Acetate
Colcolor S 35/60:
    Pigment Black: Printex 60
    Pigment Black Concentration %: 35
    Binder: Polystyrene
Colcolor SAN 25/18:
    Pigment Black: Colour Black FW 18
    Pigment Black Concentration %: 25
    Binder: Styrene acrylonitrile
Colcolor VC 45/25:
    Pigment Black: Printex 25
    Pigment Black Concentration %: 45
    Binder: Polyvinyl chloride
Colcolor VC 30/300:
    Pigment Black: Printex 300
    Pigment Black Concentration %: 30
    Binder: Polyvinyl chloride

Degussa Corp.: Pigment Black Preparations for Plastics (Continued):

Colcolor VC 35/L:
    Pigment Black: Printex L
    Pigment Black Concentration %: 35
    Binder: Polyvinyl chloride
Colcolor UNI 50:
    Pigment Black: MCF black
    Pigment Black Concentration %: 50
    Binder: Copolymer blend
Colcolor UNI 51:
    Pigment Black: MCF black
    Pigment Black Concentration %: 50
    Binder: Copolymer blend

Tack DOP 15/1:
    Pigment Black: Colour Black FW 1
    Pigment Black Concentration %: 15
    Binder: Dioctyl phthalate
Tack DOP 25/V:
    Pigment Black: Printex V
    Pigment Black Concentration %: 25
    Binder: Dioctyl phthalate
Tack DOP 30/G:
    Pigment Black: Printex G
    Pigment Black Concentration %: 30
    Binder: Dioctyl phthalate

DERUSSOL VU 25/L:
    Pigment Black: Printex L
    Pigment Black Concentration %: 25
    Binder: Water + wetting agent, anion-active
Derussol AN 1-25/L:
    Pigment Black: Printex L
    Pigment Black Concentration %: 25
    Binder: Water + wetting agent, non-ionogenic
Derussol Z 35:
    Pigment Black: Printex V
    Pigment Black Concentration %: 35
    Binder: Water + wetting agent, anion-active
Derussol 345:
    Pigment Black: Printex 300
    Pigment Black Concentration %: 45
    Binder: Water + wetting agent, anion-active

D.J. Enterprises, Inc.: Synthetic Carbon-910-44M:

Synthetic Carbon-910 is custom ground to <44 micron for uniform dispersion, ideally used in rubber compounds, coatings and plastic composites.

Easier Dispersion--Faster Wetting--Lower Cost

D.J. Enterprises offers a lower cost replacement for carbon black petroleum-based products, and provides high quality and performance.

Their product is basically derived from high purity metallurgical carbon which has been superheated to drive off organic tramp elements, then screen to <44 micron to insure easy distribution in your formulation. They are able to maintain a <1% moisture level at time of shipment, and to insure against the possibility of additional moisture contamination, they shrink-wrap all 45# multi-wall bags 50 bags/pallet, with no additional charge to customers.

Typical Chemistry Analysis:
    Carbon, Fixed: 99.0%
    Sulfur: 1.0
    Moisture: <1.0
    Spec. Gravity: 1.9-2.2

Eckart America: FLONAC Pearlescent Pigments:

Flonac pearlescent pigments belong to a versatile group of effect pigments able to create an infinite range of visual effects. From the soft sheen of a single pearl, to the shimmering iridescence of a rainbow, Flonac pearlescent pigments provide a kaleidoscope of possibilities to create three-dimensional transparent lustre and eye-catching colours.

A continuous cycle of the reflection and refraction of light through Flonac's semi-transparent pigment flakes is the secret to creating the fascinating mysterious effects and colours that capture your eye and tease the imagination.

Flonac pearlescent pigments find extensive use in the coatings, plastics and printing industries to create a palette of elegant special effects for a wide range of product applications.

The applications for Flonac pearlescent pigments are:

*Plastics	*Textiles
*Wallpaper	*Liquid Inks
*interior Coatings	

High Performance Silver Pearls:
MF 10: Pearlescent Lustre Effect: fine satin
    Light fastness: good
    Particle Size Range: 5-30 um
    TiO2 Modification: anatase
MF 11: Pearlescent Lustre Effect: fine satin
    Light fastness: very good
    Particle Size Range: 5-30 um
    TiO2 Modification: rutile
MI 10: Pearlescent Lustre Effect: brilliant
    Light fastness: good
    Particle Size Range: 10-50 um
    TiO2 Modification: anatase
MI 11: Pearlescent Lustre Effect: brilliant
    Light fastness: very good
    Particle Size Range: 10-50 um
    TiO2 Modification: rutile
MM 10: Pearlescent Lustre Effect: glittering
    Light fastness: good
    Particle Size Range: 30-120 um
    TiO2 Modification: anatase
MO 10: Pearlescent Lustre Effect: sparkling
    Light fastness: good
    Particle Size Range: 20-180 um
    TiO2 Modification: anatase
High Performance Gold Pearls:
MI 33: Pearlescent Lustre Effect: brilliant gold
    Light fastness: good
    Particle Size Range: 10-50 um
    TiO2 Modification: anatase
MO 30: Pearlescent Lustre Effect: golden yellow
    Light fastness: good
    Particle Size Range: 20-180 um
    TiO2 Modification: anatase

Eckart America: Metallic Pigments for Plastics:

STANDART Pigment Powders Aluminum:

Reflexal 120:
  Bulk density approx. kg/l: 0.5
  Cilas 715: um: 330

Reflexal 130:
  Bulk density approx. kg/l: 0.5
  Cilas 715: um 225

Reflexal 140:
  Bulk density approx. kg/l: 0.5
  Cilas 715: um: 125

Reflexal 211:
  Bulk density approx. kg/l: 0.4
  Sieving: <100 um: 99 min %

Reflexal 212:
  Bulk density approx. kg/l: 0.4
  Sieving: <71 um: 99 min %

Reflexal 214:
  Bulk density approx. kg/l: 0.4
  Sieving: <45 um: 99 min %

Lac NOT:
  Bulk density approx. kg/l: 0.4
  Sieving: <160 um: 99 min %

Lac NDT:
  Bulk density approx. kg/l: 0.3
  Sieving: <160 um: 99 min %

Chromal X:
  Bulk density approx. kg/l: 0.3
  Sieving: <45 um: 99.8 min %

Resist 211:
  Bulk density approx. kg/l: 0.3
  Sieving: <100 um: 99 min %

Resist 212:
  Bulk density approx. kg/l: 0.3
  Sieving: <71 um: 98 min %

Resist 214:
  Bulk density approx. kg/l: 0.3
  Sieving: <71 um: 98% min %

Resist 501:
  Bulk density approx. kg/l: 0.2
  Sieving: <45 um: 99 min. %

Resist 801:
  Bulk density approx. kg/l: 0.2
  Sieving: <45 um: 99 min %

STANDART Pigment Powders Bronze:

SF-15:
  Bulk density approx. kg/l: 0.9
  Sympatec D10: 30 um
  Shades: 1,2,3,4

SF-50:
  Bulk density approx. kg/l: 0.9
  Sieving: <160 um: 99 min%
  Shades: 1,2,3,4,5,6

SF-100:
  Bulk density approx. kg/l: 0.7
  Sieving: <71 um: 98 min%
  Shades: 1,2,3,4,5,6

XM-18:
  Bulk density approx. kg/l: 0.7
  Sieving: <45 um: 98 min%
  Shades: 1,2,3,4,5,6

Resist LT:
  Bulk density approx. kg/l: 0.9
  Sieving: <100 um: 99 min%
  Shades: 1,2,3,4,5,6

Resist CT:
  Bulk density approx. kg/l: 0.8
  Sieving: <71 um: 98 min%
  Shades: 1,2,3,4,5,6

Resist AT:
  Bulk density approx. kg/l: 0.7
  Sieving: <45 um: 98 min%
  Shades: 1,2,3,4

Resist Rotoflex:
  Bulk density approx. kg/l: 0.5
  Sieving: <45 um: 99 min%
  Shades: 2,3,4

Shades:
  1=Copper
  2=Pale Gold
  3=Rich Pale Gold
  4=Rich Gold
  5=English Green Gold
  6=Deep Gold

Eckart America: Metallic Pigments for Plastics (Continued):

STAPA Pigment Pastes Aluminum:
Reflexal 120: Cilas 715: D50: approx. um 330
Reflexal 130: Cilas 715: D50: approx. um 225
Reflexal 140: Cilas 715: D50: approx. um 125
Reflexal 211: Sieving: <100 um: 99 min. %
Reflexal 212: Sieving: <71 um: 99 min. %
Reflexal 214: Sieving: <45 um: 99 min. %
Reflexal 161: Sieving: <45 um: 99 min. %
Reflexal X: Sieving: <45 um: 99.8 min. %
Reflexal 40: Sieving: <45 um: 99.9 min. %

Chromal X: Sieving: <45 um: 99.8 min. %

Lac NOT: Sieving: <160 um: 99 min. %
Lac NDT: Sieving: <160 um: 99 min. %

Resist 211: Sieving: <100 um: 99 min. %
Resist 212: Sieving: <71 um: 98 min. %
Resist 214: Sieving: <71 um: 98 min. %
Resist 501: Sieving: <45 um: 99 min. %
Resist 801: Sieving: <45 um: 99 min. %
OT Perox: Sieving: <160 um: 99 min. %

STAPA Pigment Pastes Bronze:

SF-15:
    Pigment Content: 86%
    Sympatec D 90: approx. um: 130
    Shades: 1,2,3,4
SF-100:
    Pigment Content: 86%
    Sieving: <71 um: 99 min. %
    Shades: 1,2,3,4,5,6

Resist LT:
    Pigment Content: 82%
    Sieving: <100 um: 99 min. %
    Shades: 1,2,3,4,5,6
Resist AT:
    Pigment Content: 82%
    Sieving: <45 um: 98 min. %
    Shades: 1,2,3,4

SF-50:
    Pigment Content: 86%
    Sieving: <160 um: 99 min. %
    Shades: 1,2,3,4,5,6
XM-18:
    Pigment Content: 86%
    Sieving: <45 um: 98 min. %
    Shades: 1,2,3,4,5,6

Resist CT:
    Pigment Content: 82%
    Sieving: <71 um: 98 min. %
    Shades: 1,2,3,4,5,6
Resist Rotoflex:
    Pigment Content: 82%
    Sieving: <45 um: 99% min. %
    Shades: 2,3,4

Shades:
    1=Copper
    2=Pale Gold
    3=Rich Pale Gold
    4=Rich Gold
    5=English Green Gold
    6=Deep Gold

Eckart America: Metallic Pigments for Plastics (Continued):

MASTERCOLOR Aluminum Flake:
  Mastercolor are flake like aluminum pigments, which are encapsulated in silica that is colored with organic pigments.
  The Mastercolor pigments were designed to achieve multicolor effects in combination with other colorants. Particularly in darker shades their unique specular metallic shades display excellent visibility offering a rich and sophisticated appearance.
  Mastercolor pigments are available in three different shades:
  *Mastercolor Gold   125 001
  *Mastercolor Blue   125 001
  *Mastercolor Green   125 001

  All three pigments have excellent heat stability, very good light fastness and are applicable in all major plastics; including polycarbonate and acrylic resins.
  All three Mastercolor pigments have an average particle size of 125u. Presently these are only offered in powder form. However, concentrates in media indicated can be custom made at certain minimum quantities.
  All ingredients of each pigment have FDA approval.

MASTERSAFE Metal Pigment Pellets Aluminum:
05201:
    Approx. um: 5   Fineness: ultra-fine
    Principal Application: fibers, best opacity
08201:
    Approx. um: 8   Fineness: very fine
    Principal Application: Very opaque
10201:
    Approx. um: 10   Fineness: fine
    Principal Application: Very opaque
60201:
    Approx. um: 60   Fineness: Medium-coarse
    Principal Application: fine sparkle effect
125101:
    Approx. um: 125   Fineness: coarse
    Principal application: sparkle effect
225101:
    Approx. um: 225   Fineness: coarse
    Principal application: coarse sparkle effect
330101:
    Approx. um: 330   Fineness: extra coarse
    Principal application: extra coarse sparkle effect

EM Industries, Inc.: AFFLAIR Gold Pearl Lustre Pigments:

Afflair 329:
   Marathon Gold
   Microns: 5-25

Afflair 355:
   Glitter Gold
   Microns: 10-100

Afflair 309:
   Medallion Gold
   Microns: 10-60

Afflair 363:
   Shimmer Gold
   Microns: 20-150

Afflair 302:
   Satin Gold
   Microns: 5-25

Afflair 326:
   Olympic Gold Stain
   Microns: 5-25

Afflair 320:
   Bright Gold Pearl
   Microns: 10-60

Afflair 306:
   Olympic Gold
   Microns: 10-60

Afflair 300:
   Gold Pearl
   Microns: 10-60

Afflair 323:
   Royal Gold Stain
   Microns: 5-25

Afflair 351:
   Sunny Gold
   Microns: 5-100

Afflair 303:
   Royal Gold
   Microns: 10-60

AFFLAIR Earthtone Pearl Lustre Pigments

Afflair 520:
   Satin Bronze
   Microns: 5-25

Afflair 532:
   Glitter Red Brown
   Microns: 10-125

Afflair 500:
   Bronze
   Microns: 10-60

Afflair 524:
   Satin Red
   Microns: 5-25

Afflair 530:
   Glitter Bronze
   Microns: 10-125

Afflair 504:
   Red
   Microns: 10-60

Afflair 522:
   Satin Red Brown
   Microns: 5-25

Afflair 534:
   Glitter Red
   Microns: 10-125

Afflair 502:
   Red Brown
   Microns: 10-60

Afflair 600:
   Black Mica
   Microns: 10-60

EM Industries, Inc.: AFFLAIR Interference Pearl Lustre Pigments:

Afflair 211:
Fine Satin Red
Microns: 5-25

Afflair 231:
Fine Satin Green
Microns: 5-25

Afflair 215:
Rutile Red Pearl
Microns: 10-60

Afflair 235:
Rutile Green Pearl
Microns: 10-60

Afflair 259:
Flash Red
Microns: 10-125

Afflair 299:
Flash Green
Microns: 10-125

Afflair 217:
Rutile Copper Pearl
Microns: 10-60

Afflair 221:
Fine satin Blue
Microns: 5-25

Afflair 201:
Fine Satin Yellow
Microns: 5-25

Afflair 225:
Rutile Blue Pearl
Microns: 10-60

Afflair 205:
Rutile Platinum Gold
Microns: 10-60

Afflair 289:
Flash Blue
Microns: 10-125

Afflair 207:
White Gold Pearl
Microns: 5-100

Afflair 223:
Fine Satin Lilac
Microns: 5-25

Afflair 249:
Flash Gold
Microns: 10-125

Afflair 219:
Rutile Lilac Pearl
Microns: 10-60

EM Industries, Inc.: AFFLAIR Silver White Pearl Lustre Pigments:

Afflair 110:
   Fine Satin
   Microns: 1-15

Afflair 111:
   Rutile Fine Silver
   Microns: 1-15

Afflair 173:
   Silk Pearl
   Microns: 1-50

Afflair 119:
   Polar White
   Microns: 5-25

Afflair 120:
   Lustre Satin
   Microns: 5-25

Afflair 121:
   Rutile Lustre Satin
   Microns: 5-25

Afflair 123:
   Bright Lustre Satin
   Microns: 5-25

Afflair 100:
   Silver Pearl
   Microns: 10-60

Afflair 103:
   New Rutile Silver
   Microns: 10-60

Afflair 141:
   Bright Lustre Pearl
   Microns: 5-100

Afflair 151:
   Lustre Pearl
   Microns: 1-110

Afflair 153:
   Flash Pearl
   Microns: 20-100

Afflair 163:
   Shimmer Pearl
   Microns: 20-180

Afflair 183:
   Ultra Glitter
   Microns: 1-500

Engelhard Corp.: HARSHAW METEOR and METEOR PLUS for High Performance Applications:

Harshaw Meteor and Meteor Plus high temperature complex inorganic color pigments (CICP) are designed and consistently manufactured to very high quality standards. CICP's are used in high performance applications such as plastics, liquid coatings, powder coatings, and coil coatings, where very high temperature stability, chemical resistance, non-bleed/migration, permanence and long term weatherability are important.

These colorants are described as synthetic or man-made minerals. They are similar to naturally occurring minerals which have withstood the ravages of weather and eons of time unaffected by nature.

The Meteor and Meteor Plus colorants fall into two main classes of stable crystal structures: spinels and rutiles. These crystal structures provide clean, opaque and easy to disperse yellow buffs, yellow and red shade browns, blues, greens and blacks.

The manufacturing process for Meteor/Meteor Plus colorants requires high temperature chemistry. The high quality and chemically controlled raw materials are fired in a kiln or furnace at temperatures up to 1200C (2192F), where the high temperature chemistry takes place producing the highly stable spinel or rutile crystal structure. Further processing provides all the other attributes the crystal must have to satisfy pigment processing requirements.

Performance Characteristics:

Meteor and Meteor Plus pigments exhibit the following high performance characteristics:

* Superior bleed resistance in water, ethanol, diethylene glycol, xylol, toluol, butyl cellosolve, lacquer solvents, mineral spirits, paraffin, halogenated hydrocarbons, soap, or dioctyl phthalate (plasticizers).
* Excellent acid resistance.
* Excellent alkali resistance.
* Excellent color stability on baking.
* Excellent lightfastness and outdoor weatherability in both masstones and tints.
* Excellent heat resistance.
* Good dispersibility.
* High hiding power.
* Excellent warping resistance in molded olefins.

Engelhard Corp.: MEARLIN DYNACOLOR Pigments:

A series of pearlescent pigments that exhibits extremely attractive effects, especially for plastic applications, general decorative coatings and printing inks. They consist of absorption colorants deposited directly on interference pigments of titanium dioxide-coated mica. They can be used in injection molding, blow molding, extrusion and similar processes.

Dynacolor BB:
    Particle Size (u) Range: 6-48
    Blue by reflection, blue absorption

Dynacolor BG:
    Particle Size (u) Range: 6-48
    Green by reflection, blue absorption

Dynacolor BP:
    Particle Size (u) Range: 6-48
    White by reflection, blue absorption

Dynacolor BY-B:
    Particle Size (u) Range: 6-48
    Gold by reflection, blue absorption

Dynacolor GB:
    Particle Size (u) Range: 6-48
    Blue by reflection, green absorption

Dynacolor GG:
    Particle Size (u) Range: 6-48
    Green by reflection, green absorption

Dynacolor GP:
    Particle Size (u) Range: 6-48
    White by reflection, green absorption

Dynacolor GY:
    Particle Size (u) Range: 6-48
    Gold by reflection, green absorption

Dynacolor RB:
    Particle Size (u) Range: 6-48
    Blue by reflection, red absorption

Dynacolor VP:
    Particle Size (u) Range: 6-48
    White by reflection, violet absorption

Engelhard Corp.: MEARLIN HI-LITE Interference Colors:

Produce a two-color effect that cannot be produced by conventional pigments.

Hi-Lite Gold:
    Particle Size (u) Range: 6-48
    Gold by reflection, blue transmission. Very good luster, high color intensity.
Hi-Lite Orange:
    Particle Size (u) Range: 6-48
    Orange by reflection, blue-green transmission. Very good luster, high color intensity.
Hi-Lite Red:
    Particle Size (u) Range: 6-48
    Red by reflection, green transmission. Very good luster, high color intensity.
Hi-Lite Violet:
    Particle Size (u) Range: 6-48
    Violet by reflection, yellow-green transmission. Very good luster, high color intensity.
Hi-Lite Blue:
    Particle Size (u) Range: 6-48
    Blue by reflection, yellow transmission. Very good luster, high color intensity.
Hi-Lite Green:
    Particle Size (u) Range: 6-48
    Green by reflection, red transmission. Very good luster, high color intensity.
Hi-Lite Super Gold:
    Particle Size (u) Range: 6-48
    Yellow-gold by reflection, bluish transmission. Excellent luster and very intense color.
Hi-Lite Super Orange:
    Particle Size (u) Range: 6-48
    Orange by reflection, blue-green transmission. Excellent luster, good color intensity.
Hi-Lite Super Red:
    Particle Size (u) Range: 6-48
    Red by reflection, green transmission. Excellent luster and very intense color.
High Lite Super Violet:
    Particle Size (u) Range: 6-48
    Violet by reflection, yellow-green transmission. Excellent luster; good color intensity.
Hi-Lite Super Blue:
    Particle Size (u) Range: 6-48
    Blue by reflection, yellow transmission. Excellent luster and very intense color.
Hi-Lite Super Green:
    Particle Size (u) Range: 6-48
    Green by reflection, red transmission. Excellent luster and very intense color.

Engelhard Corp.: MEARLIN Micro Pearl Pigments:

Mearlin Micro Pearls are a series of luster pigments characterized by their very fine particle size. They consist of tin-free, rutile titanium dioxide and/or iron oxide coated mica. Applications include plastics incorporation, general coatings and printing inks.

Micro Gold:
    Particle Size (u) Range: 2-24
    Very fine particle size provides increased opacity and a smooth, satin-like appearance.

Micro Orange:
    Particle Size (u) Range: 2-24
    Special Properties: Same as Micro Gold

Micro Red:
    Particle Size (u) Range: 2-24
    Same as Micro Gold

Micro Violet:
    Particle Size (u) Range: 2-24
    Same as Micro Gold

Micro Blue:
    Particle Size (u) Range: 2-24
    Same as Micro Gold

Micro Green:
    Particle Size (u) Range: 2-24
    Same as Micro Gold

Micro Brass:
    Particle Size (u) Range: 2-24
    Same as Micro Gold

Micro Bronze:
    Particle Size (u) Range: 2-24
    Same as Micro Gold

Micro Copper:
    Particle Size (u) Range: 2-24
    Same as Micro Gold

Micro Russet:
    Particle Size (u) Range: 2-24
    Same as Micro Gold

Engelhard Corp.: MEARLIN "Non-Metallic," Metallic Colors:

Mearlin metallic colors produce metallic shades and effects without the use of metal flake pigments. Unlike many of the latter, Mearlin metallic pigments do not tarnish or oxidize, and are non-arcing.

Card Silver:
    Particle Size (u) Range: 6-48
    Metallic silver appearance with bright highlights.

Nu-Antique Silver:
    Particle Size (u) Range: 6-90
    Dark gunmetal color; can be blended with white Mearlins to pewter or to bright silvers.

Super Brass:
    Particle Size (u) Range: 6-48
    Similar to Brass exhibiting brighter luster and deeper color intensity.

Brass:
    Particle Size (u) Range: 6-48
    Similar to gold in a finer particle size range. Intense, brighter gold effect.

Card Gold:
    Particle Size (u) Range: 6-48
    Metallic gold appearance with bright highlights.

Sunset Gold:
    Particle Size (u) Range: 6-48
    Dark, metallic reddish-gold luster.

Majestic Gold:
    Particle Size (u) Range: 6-48
    Intense, deep, bright gold effect.

Mayan Gold:
    Particle Size (u) Range: 6-48
    Intense, deeper, bright reddish-gold effect.

Aztec Gold:
    Particle Size (u) Range: 6-50
    Intense, deep, bright orange-gold color.

Inca Gold:
    Particle Size (u) Range: 6-75
    Gold color by reflection, yellow transmission; having an intense gold effect.

Brilliant Gold:
    Particle Size (u) Range: 6-75
    Deep color by reflection, yellow transmission. Good luster, very good color intensity.

**Engelhard Corp.: MEARLIN "Non-Metallic," Metallic Colors
(Continued):**

Nu-Antique Gold:
    Particle Size (u) Range: 6-90
    Dark metallic gold color; can be blended to medium gold shades
with Inca Gold.

Sparkle Gold:
    Particle Size (u) Range: 10-110
    Similar to Inca Gold having more intense glittery gold effect.

Super Bronze:
    Particle Size (u) Range: 6-48
    Exhibiting brighter luster and intense bronze color.

Golden Bronze:
    Particle Size (u) Range: 6-90
    Bright bronze color, can be shifted toward yellow with
Brilliant Gold, and toward red with Copper.

Nu-Antique Bronze:
    Particle Size (u) Range: 6-90
    Very dark bronze color, designed to be blended with Super
Bronze for antique effects.

Super Copper:
    Particle Size (u) Range: 6-48
    Bright luster and intense copper color.

Copper:
    Particle Size (u) Range: 6-90
    Deep copper color, can be blended with Golden Bronze for
intermediate shades.

Nu-Antique Copper:
    Particle Size (u) Range: 6-90
    Very dark copper color, designed to be blended with Super
Copper for antique effects.

Super Russet:
    Particle Size (u) Range: 6-48
    Deep, russet-red having exceptional luster and superior
color intensity.

Super Red-Russet:
    Particle Size (u) Range: 6-48
    Similar to Super Russet with a redder hue.

Super Blue-Russet:
    Particle Size (u) Range: 6-48
    Similar to Super Russet with a magenta hue.

Engelhard Corp.: MEARLIN White Pearlescents:

MAGNA PEARL 3000:
    Particle Size (u) Range: 2-10
    Fine, bright white luster, higher opacity, excellent coverage

MagnaPearl 3100:
    Particle Size (u) Range: 2-10
    Rutile version of above, with increased luster

MagnaPearl 2000:
    Particle Size (u) Range: 5-25
    Smooth texture, excellent coverage

MagnaPearl 2100:
    Particle Size (u) Range: 5-25
    Rutile version of MagnaPearl 2000

MagnaPearl 2110:
    Particle Size (u) Range: 5-25
    Treated MagnaPearl 2100 for anti-yellowing

MagnaPearl 2300:
    Particle Size (u) Range: 5-25
    Bright, white luster, good coverage

Superfine:
    Particle Size (u) Range: 4-32
    Smooth texture and rich satin luster

Satin White:
    Particle Size (u) Range: 4-32
    Bright, white, cleaner luster. Excellent opacity and coverage

Supersilk:
    Particle Size (u) Range: 4-48
    High brilliance and exceptional smoothness

Superwhite:
    Particle size (u) Range: 6-48
    Bright white luster

MagnaPearl 1000:
    Particle Size (u) Range: 8-48
    Exceptional brilliance and whiteness

MagnaPearl 1100:
    Particle Size (u) Range: 8-48
    Rutile version of MagnaPearl 1000

Engelhard Corp.: MEARLIN White Pearlescents (Continued):

MagnaPearl 1110;
    Particle Size (u) Range: 8-48
    Treated MagnaPearl 1100 for anti-yellowing

Silkwhite:
    Particle Size (u) Range: 4-75
    Smooth, satin luster

Pearlwhite:
    Particle Size (u) Range: 6-90
    Good luster; most widely used and economical quality

MagnaPearl 5000:
    Particle Size (u) Range: 14-95
    Economical silvery-white sparkle

Sparkle:
    Particle Size (u) Range: 10-110
    Glittery, silver luster; quite transparent

Supersparkle:
    Particle Size (u) Range: 10-150
    Low proportions provide a high gloss "wet look" finish, higher
concentrations produce a fine sandpaper texture

MagnaPearl 4000:
    Particle Size (u) Range: 15-150
    Bright, silvery-white sparkle

Heucotech Ltd.: HEUCO Organic Pigments:

Yellow:
PY-100100:
    Form: Powder
    Pigment Type: Arylide Yellow
    C.I. Name: PY-1

PY-101402:
    Form: Powder
    Pigment Type: Diarylide Yellow AAOT
    C.I. Name: PY-14

PY-106100:
    Form: Powder
    Pigment Type: Azo Yellow
    C.I. Name: PY-61

PY-107402:
    Form: Powder
    Pigment Type: Arylide Yellow
    C.I. Name: PY-74

PY-109500:
    Form: Powder
    Pigment Type: Disazo Yellow RS
    C.I. Name: PY-95

Orange:
PO 200502:
    Form: Powder
    Pigment Type: DNA Orange
    C.I. Name: PO-5

Green:
PG 600703:
    Form: Powder
    Pigment Type: Phthalo Green
    C.I. Name: PG-7

PG 600703K:
    Form: Powder
    Pigment Type: Phthalo Green
    C.I. Name: PG-7

PG 600704:
    Form: Powder
    Pigment Type: Phthalo Green
    C.I. Name: PG-7
PG 600707:
    Form: Powder
    Pigment Type: Phthalo Green
    C.I. Name: PG-7

Heucotech Ltd.: HEUCO Organic Pigments (Continued):

Red:

PR 300202:
  Form: Powder
  Pigment Type: Naphthol Red MS
  C.I. Name: PR-2

PR 302201:
  Form: Powder
  Pigment Type: Naphthol Red YS
  C.I. Name: PR-22

PR 304830:
  Form: Powder
  Pigment Type: Strontium Red 2B
  C.I. Name: PR-48:3

PR 304910:
  Form: Powder
  Pigment Type: Barium Lithol
  C.I. Name: PR-49:1

PR 304920:
  Form: Powder
  Pigment Type: Calcium Lithol YS
  C.I. Name: PR-49:2

PR 304921:
  Form: Powder
  Pigment Type: Calcium Lithol
  C.I. Name: PR-49:2

PR 305311:
  Form: Powder
  Pigment Type: Red Lake C
  C.I. Name: PR-53:1

PR 305715:
  Form: Powder
  Pigment Type: Lithol Rubine
  C.I. Name: PR-57:1

PR 314400:
  Form: Powder
  Pigment Type: Disazo Red
  C.I. Name: PR-144

PR 316600:
  Form: Powder
  Pigment Type: Disazo Red
  C.I. Name: PR-166

PR 317000:
  Form: Powder
  Pigment Type: Naphthol Red
  C.I. Name: PR-170

Blue:

PB 501532:
  Form: Powder
  Pigment Type: Phthalo Blue
  C.I. Name: PB-15:3

PB 515303:
  Form: Powder
  Pigment Type: Phthalo Blue
  C.I. Name: PB-15:3

PB 501540:
  Form: Powder
  Pigment Type: Phthalo Blue
  C.I. Name: PB-15:4

Heucotech Ltd.: HEUCOPHOS Anticorrosive Pigments: Orthophosphate

ZP-10:
    Zinc Content as Zn, %: 51.2
    Phosphate Content as PO4, %: 47.5

ZBZ:
    Zinc Content as Zn, %: 34.5
    Phosphate Content as PO4, %: 23
    Organic Content, %: 0.2
    Silicon Content as SiO2, %: 7.3
    Magnesium Oxide, %: 3.5

ZPA:
    Zinc Content as Zn, %: 39
    Phosphate Content as PO4, %: 55.5
    Aluminum Content as Al, %: 4.5

ZPO:
    Zinc Content as Zn, %: 55.5
    Phosphate Content as PO4, %: 38.5
    Organic Content %: 0.3

ZMP:
    Zinc Content as Zn, %: 55
    Phosphate Content as PO4, %: 38.5
    Molybdate Content as MoO3, %: 1.7

ZPZ:
    Zinc Content as Zn, %: 56
    Phosphate Content as PO4, %: 38
    Organic Content, %: 0.3

ZCP:
    Zinc Content as Zn, %: 30
    Phosphate Content as PO4, %: 24
    Calcium Content as CaO, %: 14.5
    Strontium Content as SrO, %: 5
    Silicon Content as SiO2, %: 14

CHP:
    Phosphate Content as PO4, %: 67
    Calcium Content as Ca, %: 29.5

Heucotech Ltd.: HEUCOPHOS Anticorrosive Pigments: Polyphosphate:

CAPP:
    Phosphate Content as P2O5, %: 26
    Aluminum Content as Al2O3, %: 7
    Calcium Content as CaO, %: 31
    Silicon Content as SiO2, %: 28

SAPP:
    Phosphate Content as P2O5, %: 42
    Aluminum Content as Al2O3, %: 12
    Strontium Content as SrO, %: 31

SRPP:
    Phosphate Content as P2O5, %: 44
    Aluminum Content as Al2O3, %: 12
    Strontium Content as SrO, %: 28
    Magnesium Silico Fluoride, %: 0.3

ZAPP:
    Zinc Content as Zn, %: 30
    Phosphate Content as P2O5, %: 48
    Aluminum Content as Al2O3, %: 12

ZCPP:
    Zinc Content as ZnO, %: 37
    Phosphate Content as P2O5, %: 18
    Aluminum Content as Al2O3, %: 3
    Calcium Content as CaO, %: 14
    Strontium Content as SrO, %: 5
    Silicon Content as SiO2, %: 14

Heucotech Ltd.: HEUCOROX Synthetic Iron Oxide Micronized Grades:

Yellow:
Yellow 130 MF:
    Form: Powder
    Chemical Composition: $Fe_2O_3-H_2O$
    Fe_2O_3 Content, %: 85.0

Yellow 140 MF:
    Form: Powder
    Chemical Composition: $Fe_2O_3-H_2O$
    Fe_2O_3 Content, %: 85.0

Yellow 145 MF:
    Form: Powder
    Chemical Composition: $Fe_2O_3-H_2O$
    Fe_2O_3 Content, %: 85.0

Yellow 150 MF:
    Form: Powder
    Chemical Composition: $Fe_2O_3-H_2O$
    Fe_2O_3 Content, %: 85.0

Red:

Red 305 MF:
    Form: Powder
    Chemical Composition: FE_2O_3
    Fe_2O_3 Content, %: 95.0

Red 335 MF:
    Form: Powder
    Chemical Composition: Fe_2O_3
    Fe_2O_3 Content, %: 94.0

Red 310 MF:
    Form: Powder
    Chemical Composition: Fe_2O_3
    Fe_2O_3 Content, %: 95.0

Red 340 MF:
    Form: Powder
    Chemical Composition: Fe_2O_3
    Fe_2O_3 Content, %: 94.0

Red 320 MF:
    Form: Powder
    Chemical Composition: Fe_2O_3
    Fe_2O_3 Content, %: 95.0

Red 360 MF:
    Form: Powder
    Chemical Composition: Fe_2O_3
    Fe_2O_3 Content, %: 94.0

Red 330 MF:
    Form: Powder
    Chemical Composition: Fe_2O_3
    Fe_2O_3 Content, %: 95.0

Red 380 MF:
    Form: Powder
    Chemical Composition: Fe_2O_3
    Fe_2O_3 Content, %: 94.0

Red 333F:
    Form: Powder
    Chemical Composition: Fe_2O_3
    Fe_2O_3 Content, %: 92.0

Black 960 MF:
    Form: Powder
    Chemical Composition: Fe_3O_4
    Fe_3O_4 Content, %: 95.0

Heucotech Ltd.: HEUCOROX Synthetic Iron Oxides Non-Micronized
Grades:

Yellow:
Yellow 130:
    Form: Powder
    Chemical Composition: $Fe_2O_3 \cdot H_2O$
    Fe_2O_3 Content, %: 85.0

Yellow 140:
    Form: Powder
    Chemical Composition: $Fe_2O_3 \cdot H_2O$
    Fe_2O_3 Content, %: 85.0

Yellow 150:
    Form: Powder
    Chemical Composition: $Fe_2O_3 \cdot H_2O$
    Fe_2O_3 Content, %: 85.0

Red:

Red 310:
    Form: Powder
    Chemical Composition: Fe_2O_3
    Fe_2O_3 Content, %: 95.0

Red 315:
    Form: Powder
    Chemical Composition: Fe_2O_3
    Fe_2O_3 Content, %: 95.0

Red 320:
    Form: Powder
    Chemical Composition: Fe_2O_3
    Fe_2O_3 Content, %: 95.0

Red 330:
    Form: Powder
    Chemical Composition: Fe_2O_3
    Fe_2O_3 Content, %: 95.0

Red 332:
    Form: Powder
    Chemical Composition: Fe_2O_3
    Fe_2O_3 Content, %: 93.0

Red 333:
    Form: Powder
    Chemical Composition: Fe_2O_3
    Fe_2O_3 Content, %: 92.0

Red 335:
    Form: Powder
    Chemical Composition: Fe_2O_3
    Fe_2O_3 Content, %: 93.0

Red 340:
    Form: Powder
    Chemical Composition: Fe_2O_3
    Fe_2O_3 Content, %: 93.0

Red 360:
    Form: Powder
    Chemical Composition: Fe_2O_3
    Fe_2O_3 Content, %: 93.0

Red 380:
    Form: Powder
    Chemical Composition: Fe_2O_3
    Fe_2O_3 Content, %: 94

Black:
Black 960:
    Form: Powder
    Chemical Composition: Fe_3O_4
    Fe_3O_4 Content, %: 95.0

Ishihara Sangyo Kaisha, Ltd.: TIPAQUE Titanium Dioxide in Chloride Process:

CR-50:
TiO2 %: 95
Main Modifier: Al
Rank of Particle Size: M

CR-60:
TiO2 %: 95
Main Modifier: Al
Rank of Particle Size: S

CR-50-2:
TiO2 %: 95
Main Modifer: Al, Organic
Rank of Particle Size: M

CR-60-2:
TiO2 %: 95
Main Modifier: Al, Organic
Rank of Particle Size: S

CR-57:
TiO2 %: 95
Main Modifier: Al, Zr, Organic
Rank of Particle Size: M

CR-63:
TiO2 %: 97
Main Modifier: Al, Si, Organic
Rank of Particle Size: S

CR-90:
TiO2 %: 90
Main Modifier: Al, Si
Rank of Particle Size: M

CR-67:
TiO2 %: 92
Main Modifier: Al
Rank of Particle Size: S

CR-90-2:
TiO2 %: 90
Main Modifier: Al, Si, Organic
Rank of Particle Size: M

CR-58:
TiO2 %: 93
Main Modifier: Al
Rank of Particle Size: L

CR-93:
TiO2 %: 90
Main Modifier: Al, Si
Rank of Particle Size: L

CR-58-2:
TiO2 %: 93
Main Modifier: Al, Organic
Rank of Particle Size: L

CR-95:
TiO2 %: 90
Main Modifier: Al, Si, Organic
Rank of Particle Size: L

CR-85:
TiO2 %: 88
Main Modifier: Al, Si
Rank of Particle Size: M

CR-953:
TiO2 %: 90
Main Modifier: Al, Si, Organic
Rank of Particle Size: L

CR-80:
TiO2 %: 93
Main Modifier: Al, Si
Rank of Particle Size: M

CR-97:
TiO2 %: 93
Main Modifier: Al, Zr
Rank of Particle Size: M

Ishihara Sangyo Kaisha, Ltd.: TIPAQUE Titanium Dioxide in Sulphate Process:

R-820:
TiO2 %: 93
Main Modifier: Al, Si, Zn
Rank of Particle Size: M

R-830:
TiO2 %: 93
Main Modifier: Al, Si, Zn
Rank of Particle Size: M

R-930:
TiO2, %: 93
Main Modifier: Al, Zn
Rank of Particle Size: M

R-980:
TiO2, %: 93
Main Modifier: Al, Organic
Rank of Particle Size: M

R-550:
TiO2, %: 94
Main Modifier: Al, Si
Rank of Particle Size: M

R-630:
TiO2, %: 94
Main Modifier: Al
Rank of Particle Size: M

R-680:
TiO2, %: 95
Main Modifier: Al
Rank of Particle Size: S

R-670:
TiO2, %: 93
Main Modifier: Al
Rank of Particle Size: S

R-580:
TiO2, %: 94
Main Modifier: Al
Rank of Particle Size: L

R-780:
TiO2, %: 88
Main Modifier: Al, Si
Rank of Particle Size: M

R-780-2:
TiO2 %: 80
Main Modifier: Al, Si
Rank of Particle Size: M

R-850:
TiO2 %: 90
Main Modifier: Al, Si
Rank of Particle Size: M

R-855:
TiO2 %: 90
Main Modifier: Al, Si
Rank of Particle Size: M

A-100:
TiO2 %: 98
Main Modifier: -----
Rank of Particle Size: S

A-220:
TiO2 %: 96
Main Modifier: Al
Rank of particle Size: S

W-10:
TiO2 %: 98
Main Modifier: -----
Rank of Particle Size: S

Ishihara Sangyo Kaisha, Ltd.: TIPAQUE Titanium Yellow:

TY-50:
    TiO_2 %: 78
    Main Modifier: Ni, Sb
    Hue: A light color

TY-55:
    TiO_2 %: 76
    Main Modifier: Ni, Sb, Al
    Hue: A light color

TY-70:
    TiO_2 %: 78
    Main Modifier: Ni, Sb
    Hue: A dark color

TY-70S:
    TiO_2 %: 78
    Main Modifier: Ni, Sb
    Hue: A dark color

TY-100:
    TiO_2 %: 77
    Main Modifier: Cr, Sb, Al
    Hue: A light color

TY-150:
    TiO_2 %: 77
    Main Modifier: Cr, Sb
    Hue: A dark color

TY-200:
    TiO_2 %: 78
    Main Modifier: Cr, Sb
    Hue: A dark color

TY-300:
    TiO_2 %: 73
    Main Modifier: Cr, Sb
    Hue: A dark color

Kemira Pigments, Inc.: KEMIRA 460 Rutile Titanium Dioxide:

Kemira 460 is a rutile titanium dioxide pigment designed with high strength, very low oil absorption, with the clean, blue tint tone commonly associated with anatase titanium dioxide. It has excellent dispersibility and shows good resistance to antioxidant yellowing in polyolefins. Kemira 460 rutile titanium dioxide meets the requirements of specification ASTM D 476-84 (1989), Type II & III.

Suggested Applications:
* Plastics: vinyl, polyolefins, ABS, polystyrene, acrylics, plastic pipe
* Floor Coverings: vinyl tile, asphalt tile
* Rubber: tile, latex, compounding

Product Specification:

Property:	Range:
Color Brightness, CIE L Value	96.5 Minimum
Color Tone, CIE b value	0.9-1.9
Tinting Strength	1500 Minimum
pH	6.0-9.0
+325 Mesh Residue (brushed screen)	0.05% Maximum
Oil Absorption	75-125% of Standard
Specific Resistance	6000 ohm/cm Minimum
Water Soluble Salts, as Na2SO4	1000 ppm

Additional Properties (Typical):
TiO2: 97%
P2O5: 0.10%
SiO2: 0.20%
Al2O3: 1.50% Maximum
Rutile: 99%
LOI: 0.80%
H2O: 0.3% Maximum
Particle Size: 0.17 microns
Oil Absorption: 16 grams oil/100 grams pigment

Additional Information:
* Kemira 460 contains no Class I or Class II Ozone Depleting Substances
* Kemira 460 meets the CONEG Certification for Heavy Metal Content of less than 100 ppm
* Kemira 460 is covered under Title 21 Part 174 of the Food and Drug Cosmetic Act
Specific areas include:
175.300 Resinous and Polymeric Coatings
176.170 Components of paper and paperboard in contact with aqueous and fatty foods
176.180 Components of paper and paperboard in contact with dry foods
178.3297C Colorants for polymers
Kemira 460 complies with NSF Standard 14

Kemira Pigments Inc.: KEMIRA 470 Rutile Titanium Dioxide:

Kemira 470 is a rutile titanium dioxide pigment designed to meet the needs of the most demanding plastics applications. It has the same brightness and tone properties as standard Kemira 460 but is end-treated to impart hydrophobic properties. This results in outstanding dispersibility and significantly improved high temperature performance. Kemira 470 rutile titanium dioxide meets the requirements of specification ASTM D 476-84 (1989), Type II & III.

Suggested Applications:
* Thin film where dispersion is critical
* High temperature thermoplastics
* General purpose concentrates
* Rigid and flexible vinyl products

Product Specification:

Property:	Range:
Color Brightness, CIE L Value	96.5 Minimum
Color Tone, CIE b Value	0.7-1.7
Tinting Strength	1500 Minimum
Oil Absorption	75-125% of standard

Additional Properties (Typical):
TiO_2: 97%
P_2O_5: 0.10%
SiO_2: 0.40%
Al_2O_3: 1.50% Maximum
Rutile: 99%
LOI: 0.55%
H_2O: 0.15% Maximum
Particle Size: 0.17 microns
Oil Absorption: 16 grams oil/100 grams pigment

Additional Information:
* Kemira 470 contains no Class I or Class II Ozone Depleting Substances
* Kemira 470 meets the CONEG Certification for Heavy Metals Content of less than 100 ppm
* Kemira 470 is covered under Title 21 Part 174 of the Food and Drug Cosmetic Act
  Specific areas include:
  175.300 Resinous and Polymeric Coatings
  176.170 Components of paper and paperboard in contact with aqueous and fatty foods
  176.180 Components of paper and paperboard in contact with dry foods
  176.3297 Colorants for polymers
  Kemira 470 complies with NSF Standard 14

Kemira Pigments Inc.: KEMIRA 630 Rutile Titanium Dioxide:

Kemira 630 is a zirconium-treated rutile titanium dioxide pigment designed to give excellent exterior durability, high chalk resistance, excellent gloss retention, high tinting strength, and high color brightness. It has a clean, blue tint tone. Kemira 630 meets the requirements of specification ASTM D 476-84 (1989), Type III & IV.

Suggested Applications:
* Surface coatings: industrial finishes, automotive coatings, coil coatings, lacquers, architectural finishes, latex paints and water-borne coatings over a wide PVC range
* Plastics: vinyl sheeting, plastisols, plastic pipe

Product Specification:

Property:	Range:
Color Brightness, "L" Value	96.5 Minimum
Color Tone, "b" value	0.2-1.4
Tinting Strength	1600 Minimum
pH	6.0-9.0
Oil Absorption	75-125% of Standard
Specific Resistance	3000 ohm/cm Minimum
Water Soluble Salts, as Na2SO4	2120 ppm

Additional Properties (Typical):
TiO2: 92%
P2O5: 0.35%
SiO2: 0.20%
Al2O3: 4%
Zr: 0.40%
LOI: 1.80%
H2O: 1.00%
Rutile: 99.5%
Particle Size: 0.2 microns
Oil Absorption: 25 grams oil/100 grams pigment

Additional Information:
* Kemira 620 contains no Class I or Class II Ozone Depleting Substances
* Kemira 620 meets the CONEG Certification for Heavy Metal Content of less than 100 ppm
* Kemira 620 is covered under Title 21 Part 174 of the Food and Drug Cosmetic Act
Specific areas include:
175.300 Resinous and Polymeric Coatings
176.170 Components of paper and paperboard in contact with aqueous and fatty foods
176.180 Components of paper and paperboard in contact with dry foods
178.3297 Colorants for polymers

Kerr-McGee Chemical Corp.: TRONOX CR-834 Chloride Process Titanium Dioxide:

Tronox CR-834 is an alumina stabilized, chloride process rutile pigment designed for a wide range of plastics and rubber applications. It is widely used in general purpose polyolefin masterbatches; polystyrene; rigid PVC pipe, fittings, side and profile formulations; flexible PVC for calendering; molding; extrusion and plastisols, and TPR and EVA for shoe applications.

CR-834 features excellent brightness and color with a very high tint strength. The blue undertone and high tint strength of CR-834 make it very desirable for resins which tend to yellowness, such as ABS.

CR-834 complies with the Plastics Pipe Institute guidelines for formulation substitution and is authorized by the National Sanitation Foundation International for interchangeability in NSF listed formulations.

Typical Properties:
    TiO2 Content, %: 97
    Density:
        Specific Gravity: 4.2
        Bulk, lb/cu ft: 40-55
    Oil Absorption: 15
    pH: 6.7
    Specific Resistance, ohms: 7,000
    Average Particle Size, um: 0.17
    Durability: fair
    Specification: ASTM D-476-84, Type II
                   ISO 591-85 R1

Kronos Inc.: KRONOS Titanium Dioxide Pigments in Plastics-Properties:

Kronos 1001:
   * has good all-round pigment properties
   * is suitable for pigmenting materials on which no demands
     are placed in terms of weather resistance, such as phen-
     olic moulding compounds and elastomers

Kronos 1075:
   * characterised by its low abrasion effect in pigmented
     plastics
   * imparts a neutral undertone to white plastics
   * leads the field among the anatase grades in terms of
     lightening and hiding power

Kronos 2073:
   * prevents lacing in extrusion coating of paper with PE
   * is especially suitable for plastics processed at high
     temperatures
   * confers high brightness and a neutral undertone in white
     plastics

Kronos 2081:
   * is a special pigment for use in plastics when the highest
     degree of weather resistance is demanded
   * is easily dispersible
   * imparts outstanding lightfastness to urea/melamine moulding
     compounds

Kronos 2200:
   * is a low-stabilised rutile grade with very good optical
     properties
   * shows very little tendency to dusting
   * is very economical in linoleum, PVC flooring, rubber and
     pigment preparations
   * confers good lightfastness in plastics used indoors

Kronos 2210:
   * is readily wetted and dispersed
   * can be used to manufacture concentrates with very high
     pigment contents
   * produces high brightness with only a slight yellowish
     undertone in white plastics

**Kronos, Inc.: KRONOS Titanium Dioxide Pigments in Plastics-
Properties (Continued):**

Kronos 2220:
   * is the leading Kronos pigment for plastics
   * confers maximum brightness and a neutral undertone in
     white plastics
   * has high lightening and hiding power
   * imparts high weather resistance to plastics used outdoors
   * meets highest requirements in terms of dispersibility

Kronos 2222:
   * satisfies highest demands on weather resistance
   * inmparts brilliance to coloured plastics
   * has superior lightening and hiding power
   * is very readily dispersible

Kronos 2230:
   * prevents depolymerisation of plastics, even at high temp-
     eratures
   * imparts outstanding optical properties and good weather
     resistance to engineering plastics
   * is readily dispersible

Kronos 2257:
   * is readily dispersible
   * imparts high weather resistance
   * is intended for plastics in which a yellowish undertone
     is desired

Kronos 2300:
   * has outstanding optical properties
   * imparts brilliance to coloured plastics
   * has very high hiding power
   * is readily incorporated in PVC plastisols, casting resins
     and elastomers

Millenium Inorganic Chemicals: TIONA Titanium Dioxide Pigments for Plastics:

Tiona RCL-188:

A next generation titanium dioxide pigment for the plastics industry that combines excellent processing performance with a strong, clean, blue tone. The proprietary surface treatment has been optimized to provide rapid wetting and easy incorpora tion into any resin system with minimal effect on melt flow properties. Tiona RCL-188 resists lacing and die lip build-up which makes it particularly suitable for highly loaded polyolefin concentrates for film and extrusion coating applications. The combination of easy incorporation and excellent performance offers cost savings potential for compounders and processors. Tiona RCL-188's outstanding color and dispersion performance in many resins also makes it an excellent candidate for use in color critical applications.

Type: Rutile chloride process titanium dioxide
Surface Treatment: Phosphate, Organic
Typical Properties:
TiO2, %: 97
Dispersion: Superior
Tint Strength: Very High
Tint Tone: Bluest
Exterior Durability: Medium
Processing Rheology: Superior
Moisture, % as Packed: <0.3

Tiona RCL-4:

A proven titanium dioxide product for the plastics industry widely used in polyolefin concentrates where its proprietary surface treatment provides excellent dispersion and retention of rheological performance. Tiona RCL-4 resists the yellowing reaction that occurs in polyolefins with certain phenolic antioxidants. It is also recommended for use in non-durable rigid and flexible PVC applications. The consistent tint tone and high tint strength of Tiona RCL-4 makes it a preferred product for custom color applications. Tiona RCL-4 is NSF listed for plastic pipe applications.

Type: Rutile chloride process titanium dioxide
Surface Treatment: Alumina, Organic
Typical Properties:
TiO2, %: 97
Dispersion: Superior
Tint Strength: Very High
Tint Tone: Blue
Exterior Durability: Medium
Processing Rheology: Excellent
Moisture, % as Packed: <0.3

Millenium Inorganic Chemicals: TIONA Titanium Dioxide Pigments for Plastics (Continued):

Tiona RCL-6:
   A durable titanium dioxide product that provides maximum color stability and chalking resistance in exterior applications. The pigment particles of Tiona RCL-6 are fully encapsulated with silica which minimizes chalking. Tiona RCL-6 is the product of choice for tinted vinyl siding and window profile systems where exterior durability is critical. Tiona RCL-6 is highly recommended in other applications where color stability is critical.
   Type: Rutile Chloride Process Titanium Dioxide
   Surface Treatment: Alumina, Silica
Typical Properties:
   TiO_2, %: 88
   Dispersion: Fair
   Tint Strength: Fair
   Tint Tone: Yellow
   Exterior Durability: Highest
   Processing Rheology: Fair
   Moisture, % as Packed: <1%

Sun Chemical Corp.: Pigments Recommended for Plastic & Rubber:

Pigment Red 38 21120:
   Pyrazolone Red
   Major Use

Pigment Red 48:1 15865:1:
   Barium 2B
   Major Use

Pigment Red 48:2 15865:2:
   Calcium 2B
   Major Use

Pigment Red 48:3 15865:3:
   Strontium Red 2B
   Some Use

Pigment Red 49:1 15630:2:
   Barium Lithol
   Some Use

Pigment Red 49:2 15630:2:
   Calcium Lithol
   Some Use

Pigment Red 53:1 15585:1:
   Red Lake C
   Some Use

Pigment Red 57:1 15850:1:
   Lithol Rubine
   Major Use

Pigment Red 60:1 16105:1:
   Pigment Scarlet
   Major Use

Pigment Red 63:1 15880:1:
   BON Maroon
   Some Use

Pigment Red 122 73915:
   Quinacridone Magenta
   Major Use

Pigment Red 170 12475:
   Naphthol Red
   Major Use

Sun Chemical Corp.: Pigments Recommended for Plastic & Rubber (Continued):

Pigment Red 179 71130:
  Perylene Maroon
  Major Use

Pigment Violet 19 73900:
  Quinacridone Red
  Major Use

Pigment Violet 19 73900:
  Quinacridone Violet
  Major Use

Pigment Violet 23 51319:
  Carbazole Violet
  Major Use

Pigment Violet 29 71129:
  Perylene Violet
  Major Use

Pigment Blue 15/15:1/15:2 74160:
  Red Alpha Shade Phthalocyanine Blue
  PB 15: Some Use
  PB 15:1: Major Use
  PB 15:2: Some Use

Pigment Blue 15:3/15:4:
  Beta Green Shade Phthalocyanine Blue GS
  PB 15:3: Major Use
  PB 15:4: Major Use

Pigment Green 7 74260:
  Phthalocyanine Green Blue Shade
  Major Use

Pigment Green 36 74265:
  Phthalocyanine Green Yellow Shade
  Major Use

Pigment Yellow 62 13940:
  Arylide Yellow Calcium Salt
  Major Use

Pigment Yellow 73 11738:
  Arylide Yellow
  Major Use

**Sun Chemical Corp.: Pigments Recommended for Plastic & Rubber
(Continued):**

Pigment Yellow 12 21090:
  Diarylide AAA Yellow
  Some Use

Pigment Yellow 13 21100:
  Diarylide AAMX Yellow/RS
  Some Use

Pigment Yellow 14 21095:
  Dairylide AAOT Yellow
  Major Use

Pigment Yellow 17 21105:
  Diarylide AADA Yellow
  Some Use

Pigment Yellow 83 21108:
  Dairylide AADMC Yellow
  Major Use

Pigment Orange 13 2110:
  Pyrazolone or Diarylide Orange
  Some Use

Pigment Orange 16 21160:
  Dianisidine Orange
  Some Use

Pigment Orange 36 11780:
  Benzimidazolone Orange
  Some Use

Pigment Orange 46 15062:
  CLARION Red
  Some Use

Section XV
Plasticizers and Esters

Akcros Chemicals: LANKROFLEX Epoxy Plasticizers:

Lankroflex L: Epoxidised Linseed Oil
Lankroflex E2307: Epoxidised Soya Bean Oil
Lankroflex E2414: Epoxidised Soya Bean Oil
Lankroflex ED6: Octyl Epoxy Stearate

Lankroflex L:
   is the most powerful thermal co-stabilizer of the epoxies
on the range, and is recommended for the most critical
applications. It has certain food approvals, and is often
preferred in medical products, because of the lower dose
level necessary to attain a given level of stabilisation.
It confers excellent resistance to extraction by water, oils
and solvents, and has a much higher viscosity than the other
types of epoxy.

Lankroflex E2314:
   is a low iodine value grade of epoxy soya bean oil giving
greater resistance to exudation and better control of colour
in sensitive food contact applications.

Lankroflex E2307:
   is a low odour epoxy plasticizer, with worldwide food
contact and medical approvals. It can be used in all types of
PVC formulations (both flexible and rigid) in combination
with metal soap stabilizers, particularly the relatively
weak calcium zinc stabilizers used in non-toxic applications.
Lankroflex E2307 has excellent resistance to extraction by
water, oils and solvents.

Lankroflex ED6:
   This epoxy ester is a good low temperature plasticizer and
effective co-stabilizer in flexible PVC formulations. Its
low viscosity and low solvation action are particularly bene-
ficial in plastisol formulations to improve their rheology
characteristics, both as manufactured and after extended
storage. Lankroflex ED6 possesses an advantage over other
epoxy plasticizers in that it does not winterize at low
temperatures, and therefore is more easily handled in cold
climates.

BP Amoco Chemicals: AMOCO Polybutenes:

Amoco polybutenes are a family of viscous, non-drying liquid polymers. They are colorless, virtually odorless, chemically stable and resist oxidation by light and moderate heat. Their unique characteristics can be used to enhance and improve the performance properties of a wide variety of end-products in many diferent industries.

Specifications:

L-14:
    Viscosity, Kinematic, ASTM D445: at 38C (100F), cSt: 27-33
    Flash Point, C (F), min: Cleveland Open Cup: 138 (280)
    Specific gravity, at 15.5C (60F): 0.830-0.845
    Color, APHA, max Photometric, haze-free: 70
    Haze, Photometric, max.: 15
    Appearance: clear

L-50:
    Viscosity, Kinematic, ASTM D445: at 38C (100F), cSt: 106-112
    Flash point, C (F), min: Cleveland Open Cup: 138 (280)
    Specific gravity, at 15.5C (60F): 0.845-0.860
    Color, APHA, max Photometric, haze-free: 70
    Haze, Photometric, max: 15
    Appearance: clear

L-65:
    Viscosity, Kinematic, ASTM D445: at 38C (100F), cSt: 116-128
    Flash point, C(F), min: Cleveland Open Cup: 149 (300)
    Specific gravity, at 15.5C (60F): 0.845-0.860
    Color, APHA, max Photometric, haze-free: 70
    Haze, Photometric, max: 15
    Appearance: clear

L-100:
    Viscosity, Kinematic, ASTM D445: at 38C (100F), cSt: 210-227
    Flash Point, C(F), min: Cleveland Open Cup: 141 (285)
    Specific gravity, at 15.5C (60F): 0.850-0.865
    Color, APHA, max Photometric, haze-free: 70
    Haze, Photometric, max: 15
    Appearance: clear

H-15:
    Viscosity, Kinematic, ASTM D445: at 99C (210F), cSt: 29-35
    Flash Point, C(F), min: Cleveland Open Cup: 141 (285)
    Specific gravity, at 15.5C (60F): 0.860-0.871
    Color, APHA, max Photometric, haze-free: 70
    Haze, Photometric, max: 15
    Appearance: clear

BP Amoco Chemicals: AMOCO Polybutenes (Continued):

Specifications:

H-25:
 Viscosity, Kinematic, ASTM D445, at 99C (210F), cSt: 48-56
 Flash Point, C(F), min. Cleveland Open Cup: 149 (300)
 Specific gravity, at 15.5C (60F): 0.868-0.879
 Color, APHA, max. Photometric, haze-free: 70
 Haze, Photometric, max.: 15
 Appearance: clear

H-35:
 Viscosity, Kinematic, ASTM D445, at 99C (210F), cSt: 73-81
 Flash Point, C(F), min. Clevelend Open Cup: 154 (310)
 Specific gravity, at 15.5C (60F): 0.871-0.887
 Color, APHA, max. Photometric, haze-free: 70
 Haze, Photometric, max: 15
 Appearance: clear

H-40:
 Viscosity, Kinematic, ASTM D445, at 99C (210F), cSt: 74-95
 Flash Point, C(F), min. Cleveland Open Cup: 163 (325)
 Specific gravity, at 15.5C (60F): 0.875-0.890
 Color, APHA, max. Photometric, haze-free: 70
 Haze, Photometric, max: 15
 Appearance: clear

H-50:
 Viscosity, Kinematic, ASTM D445, at 99C (210F), cSt: 109-125
 Flash point, C(F), min. Cleveland Open Cup: 154 (310)
 Specific gravity, at 15.5C (60F): 0.876-0.893
 Color, APHA, max. Photometric, haze-free: 70
 Haze, Photometric, max: 15
 Appearance: clear

H-100:
 Viscosity, Kinematic, ASTM D445, at 99C (210F), cSt: 196-233
 Flash point, C(F), min. Pensky Martens Closed Cup: 150(311)
 Specific gravity, at 15.5C (60F): 0.885-0.902
 Color, APHA, max. Photometric, haze-free: 50
 Haze, Photometric, max: 12
 Appearance: clear

H-300:
 Viscosity, Kinematic, ASTM D445, at 99C (210F), cSt: 635-690
 Flash Point, C(F), min. Pensky-Martens Closed Cup: 160(320)
 Specific gravity, at 15.5C (60F): 0.893-0.910
 Color, APHA, max. Photometric, haze-free: 50
 Haze, Photometric, max: 12
 Appearance: clear

BP Amoco Chemicals: AMOCO Polybutenes (Continued):

H-1500:
   Viscosity, Kinematic, ASTM D445 at 99C (210F), cSt:
3,026-3,381
     Flash Point, C(F), min.: Pensky-Martens Closed Cup: 170(338)
     Specific gravity, at 15.5C (60F): 0.896-0.913
     Color, APHA, max. Photometric, haze-free: 50
     Haze, Photometric, max.: 12
     Appearance: clear

H-1900:
   Viscosity, Kinematic, ASTM D445 at 99C (210F), cSt:
4,069-4,382
     Flash point, C(F), min.: Pensky-Martens Closed Cup: 170 (338)
     Specific gravity, at 15.5C (60F): 0.900-0.917
     Color, APHA, max., Photometric, haze-free: 50
     Haze, Photometric, max.: 12
     Appearance: clear

Aristech Chemical Corp.: ARISTECH Plasticizers:

PX-111 Diundecyl Phthalate (DUP):
 A primary plasticizer developed for wire and cable insulation compounds. PX-111 has a higher degree of linearity than many commercial DUP's and thus shows superior performance in effic-iency, processing and low temperature flexibility.
 Formulations made from PX-111 have low volatility and excell-ent resistance to chemical breakdown at high temperatures; and therefore, better retention of properties after oven aging. It can be blended with other linear phthalates.
Physical Properties:
  Molecular Weight: 474
  Specific Gravity, 25/25C: 0.952
  Boiling Point at 3.5 mm Hg, C: 262
  Pour Point, C: 9
  Flash Point, F, C.O.C.: 490
  Viscosity, 25C, cps: 52
  Refractive Index, 25C: 1.481
Specifications:
  Color, A.P.H.A., Max.: 50
  Acidity (% as Acetic Acid), Max.: 0.007
  Ester Content, % Min.: 99.6
  Moisture, % H2O, Max.: 0.1
  Odor: Characteristic

PX-120 Diisodecyl Phthalate (DIDP):
 Developed to fulfill the need for an economical vinyl plastic-izer in formulations which require low volatility and good per-formance characteristics. The low volatility of PX-120 makes it particularly outstanding for use in high temperature processing cycles, especially in those involving exposure of large surface areas.
 Vinyl products containing PX-120 are characterized by long life and a high degree of resistance to extraction by various materials.
Physical Properties:
  Molecular Weight: 447
  Specific Gravity, 25/25C: 0.965
  Boiling Point at 3.5 mm Hg, C: 255
  Pour Point, C: -48
  Flash Point, F, C.O.C.: 450
  Viscosity, 25C, cps: 86
  Refractive Index, 25C: 1.484
Specifications:
  Color, A.P.H.A., Max.: 25
  Acidity (% as Acetic Acid), Max.: 0.007
  Ester Content, % Min.: 99.6
  Moisture, % H2O, Max.: 0.1
  Odor: Characteristic

Aristech Chemical Co.: ARISTECH Plasticizers (Continued):

PX-138 Di-2 Ethyl Hexyl Phthalate (DEHP):*
  The most widely used of all vinyl plasticizers. It was one of the first plasticizers to be used with vinyl resin and remains the standard of comparison for other vinyl resin plasticizers.
  In vinyl resins, Di-2 Ethyl Hexyl Phthalate exhibits an excellent combination of overall properties, such as good low temperature performance, low extraction by oil, water, and soap solutions, as well as high clarity, good heat and light stability, and permanent flexibility.
Physical Properties:
  Molecular Weight: 390
  Specific Gravity, 25/25C: 0.983
  Boiling Point at 3.5 mm Hg, C: 230
  Pour Point, C: below -50
  Flash Point, F, C.O.C.: 420
  Viscosity, 25C, cps: 57
  Refractive Index, 25C: 1.485
Specifications:
  Color, A.P.H.A., Max.: 25
  Acidity (% as Acetic Acid), Max.: 0.007
  Ester Content, % Min.: 99.6
  Moisture, % H2O, Max.: 0.1
  Odor: Characteristic
* Industry has commonly referred to this product as Dioctyl
  Phthalate

PX-139 Diisononyl Phthalate (DINP):
  An isononyl phthalate produced from a narrow isomer distribution alcohol. Compared to highly branched isononyl phthalate, this plasticizer is more linear in nature providing measurable advantages in low temperature performance, volatility and efficiency. Dispersion resin processors will encounter lower viscosity and improved viscosity stability than in plastisols utilizing traditional DINP, DIOP and DHEP. The low volatility of PX-139 produces plastisols with reduced fuming during processing which may allow for DIDP replacement with some gain in processability.
Physical Properties:
  Molecular Weight: 418
  Boiling Point at 3.5mm Hg, C: 255
  Freeze Point, C: -45
  Flash Point, F, C.O.C.: 470
  Viscosity, 25C, cps: 70
  Refractive Index, 25C: 1.484
Specifications:
  Color, A.P.H.A., Max.: 25
  Acidity (% as Acetic Acid), Max.: 0.007
  Ester Content, % Min.: 99.6
  Moisture, % H2O, Max.: 0.1
  Odor: Characteristic

Aristech Chemical Co.: ARISTECH Plasticizers (Continued):

PX-238 Di-2 Ethyl Hexyl Adipate (DEHA):*
Widely used with polyvinyl chloride resins as a highly efficient plasticizer that will impart excellent low temperature
properties at moderate cost as well as excellent heat stability
and weathering resistance. It is extensively used in the manufacture of calendered sheeting where its good hand, drape,
clarity, and "dry surface" result in increased sales appeal to
the consumer. It is also used in extrusion compounds and in
some film to impart improved low temperature flexibility.
Physical Properties:
    Molecular Weight: 371
    Specific Gravity, 25/25C: 0.925
    Boiling Point at 3.5mm Hg, C: 214
    Freeze Point, C: -42
    Flash Point, F, C.O.C.: 390
    Viscosity, 25C, cps: 11
    Refractive Index, 25C: 1.446
Specifications:
    Color, A.P.H.A., Max.: 25
    Acidity (% as Acetic Acid), Max.: 0.01
    Ester Content, % Min.: 99.6
    Moisture, % H2O, Max.: 0.1
    Odor: Characteristic
* Industry has commonly referred to this product as Dioctyl
  Adipate

PX-239 Diisononyl Adipate (DINA):
Offers a combination of low specific gravity, high performance, low volatility and excellent low temperature properties in
polyvinyl chloride formulations. Heat and light stability of
materials plasticized with PX-239 are comparable to Di-2 Ethyl
Hexyl Adipate. PX-239 is the adipate of choice when service
conditions suggest that DEHA is too volatile and the compatibility of DIDA is too low.
Physical Properties:
    Molecular Weight: 398
    Specific Gravity, 25/25C: 0.921
    Boiling Point at 3.5 mm Hg, C: 437
    Freeze Point, C: -65
    Flash Point, F, C.O.C.: 445
    Viscosity, 25C, cps: 18
    Refractive Index, 25C: 1.448
Specifications:
    Color, A.P.H.A., Max.: 25
    Acidity (% as Acetic Acid), Max.: 0.010
    Ester Content, % Min.: 99.6
    Moisture, % H2O, Max.: 0.1
    Odor: Characteristic

Aristech Chemical Co.: ARISTECH Plasticizers (Continued):

PX-306 Di Normal Hexyl Phthalate (DNHP):
    A low-molecular-weight linear phthalate designed to perform
as a primary plasticizer and a PVC processing aid. It is
designed to compete with the branched phthalates DHP and DIHP,
BBP and benzoate plasticizers.
    PX-306 contributes all the processing advantages of low-
molecular-weight branched phthalate plasticizers while adding
the benefits of linears:
    * Improved low temperature performance
    * Reduced process fumes
    * Improved product aging
    * Lower early viscosity
Physical Properties:
    Molecular weight: 334
    Specific gravity, 25/25C: 1.003
    Boiling point at 5 mm Hg, C: 211
    Pour Point, C: below -50
    Flash point, F, C.O.C.: 380
    Viscosity, 25C, cps: 22
    Refractive Index, 25C: 1.485
Specifications:
    Color, A.P.H.A., max.: 25
    Acidity (% as acetic acid), max.: 0.007
    Ester Content, % min.: 99.6
    Moisture, % H2O, max.: 0.1
    Odor: Characteristic

PX-316 Mixed Normal Alkyl Phthalate (6-10 Phthalate):
    An efficient primary plasticizer for polyvinyl chloride
resins and copolymers.
    The 100% linearity of PX-316 imparts superior low temp-
erature flexibility to vinyl compositions, and demonstrates
excellent light and heat stability, as well as low volatility.
Addition of PX-316 to plastisols and organosols lowers initial
viscosity and leads to longer shelf life.
    PX-316 is recommended for use in sheeting, film, extrusions,
and dispersions where low temperature properties are important
and where a good compatibility and good volatility characteristics are required. It also imparts good hand, drape and clarity.
Physical Properties:
    Molecular Weight: 415
    Specific Gravity, 25/25C: 0.970
    Freeze Point, C: -20
    Flash Point, F, C.O.C.: 440
    Viscosity, 25C, cps: 34
    Refractive Index, 25C: 1.482
Specifications:
    Color, A.P.H.A., Max.: 25
    Acidity (% as Acetic Acid), Max.: 0.007
    Ester Content, % Min.: 99.6
    Moisture, % H2O, Max.: 0.1

Aristech Chemical Co.: ARISTECH Plasticizers (Continued):

PX-336 Mixed Normal Alkyl Trimellitate (6-10 TM):
   A primary plasticizer for polyvinyl chloride resins and copolymers.
   When compared to TOTM and TINTM, formulations made with PX-336 exhibit superior low temperature flexibility and resistance to oxidative degradation at high temperatures. This trimellitate offers a unique combination of easy processability, a high degree of permanence and good compatibility.
Physical Properties:
   Molecular Weight: 585
   Specific Gravity, 25/25C: 0.975
   Boiling Point at 3.5mm Hg, C: 275
   Freeze Point, C: -17
   Flash Point, F C.O.C.: 532
   Viscosity, 25C, cps: 103
   Refractive Index, 25C: 1.482
Specifications:
   Color, A.P.H.A., Max.: 200
   Acidity (% as Acetic Acid), Max.: 0.010
   Ester Content, % Min.: 99.0
   Moisture, % H2O, Max.: 0.1
   Odor: Characteristic

PX-338 Trioctyl Trimellitate (TOTM):
   A primary, monomeric plasticizer exhibiting excellent permanence for use in polyvinyl chloride resins and copolymers.
   When used as the sole plasticizer with polyvinyl chloride, PX-338 offers a degree of permanence comparable to polymeric plasticizers. At the same time its ease of processing, compatibility and low temperature properties approach those of conventional phthalates. Blends of polymeric plasticizer and PX-338 find
many cost effective applications.
Physical Properties:
   Molecular Weight: 546
   Specific Gravity, 25/25C: 0.988
   Boiling Point at 3.5mm Hg, C: 260
   Pour Point, C: -46
   Flash Point, F, C.O.C.: 490
   Viscosity, 25C, cps: 216
   Refractive Index, 25C: 1.485
Specifications:
   Color, A.P.H.A., Max.: 200
   Acidity (% as Acetic Acid), Max.: 0.02
   Ester Content, % Min.: 99.0
   Moisture, % H2O, Max.: 0.1
   Odor: Characteristic

Aristech Chemical Co.: ARISTECH Plasticizers (Continued):

PX-911 Mixed Alkyl Phthalate:
  A linear C9, C10, C11 phthalate made from normal alcohol,
is a primary plasticizer for polyvinyl chloride and copolymer
resins.
  PX-911 offers excellent permanence, low volatility, good
efficiency, and good retention of physical properties for heat
aging vinyl applications. Heat and light stability of PX-911
is superior to phthalate esters made from branched chain
alcohols.
  PX-911 is suggested for use in sheeting, film extrusion,
and dispersions where good low temperature properties are
required and where low volatility insures a greater permanence.
It is especially recommended in automotive applications for
low fog properties and for wire insulation.
  PX-911 is a medium molecular weight linear phthalate.
Physical Properties:
  Molecular Weight: 450
  Specific Gravity, 25/25C: 0.959
  Freeze Point, C: -20
  Flash Point, F, C.O.C.: 440
  Viscosity, 25C, cps: 44
  Refractive Index, 25C: 1.481
Specifications:
  Color, A.P.H.A., Max.: 50
  Acidity (% as Acetic Acid), Max.: 0.007
  Ester Content, % Min.: 99.6
  Moisture, % H2O, Max.: 0.1
  Odor: Characteristic

Arizona Chemical Co.: UNIFLEX Monomeric Plasticizers:

All Uniflex monomeric plasticizers provide the following benefits to polymer and rubber formulations:
* Low temperature flexibility
* Low viscosity, allowing ease of handling
* Excellent viscosity stability in PVC based plastisols

Uniflex BYO:
Light Liquid
Primary plasticizer and processing aid for natural and synthetic rubbers

BYS-TECH:
Colorless liquid
Plasticizer for natural and synthetic rubbers

192:
Yellow liquid
Plasticizer for synthetic rubbers

DBS:
Colorless liquid
FDA approved for PVDC food wrap and for nitrile and neoprene rubbers

DOS:
Light Liquid
Low temperature plasticizer for PVC and various chlorinated and nitrile rubbers

DCS:
Pale yellow liquid
Plasticizer for PVC film and various synthetic rubbers

301:
Light liquid
PVC applications which require low temperature flexibility and permanence

307:
Light liquid
PVC applications which require low temperature flexibility and permanence

DCP:
Light liquid
Non-toxic plasticizer for PVC (DOP replacement)

143:
Colorless liquid
Plasticizer for PVC film and plastisols

DOA:
Colorless liquid
Low temperature plasticizer for PVC film, FDA approved for cellophane

Arizona Chemical Co.: UNIFLEX Polymeric Plasticizers:

All Uniflex polymeric plasticizers provide the following benefits to polymer and rubber formulations:
* Excellent permanence, lower volatility
* Improved migration resistance
* Good low temperature flexibility
* Improved resistance to extraction in solvents, ie., gasoline, soaps, detergents, water

Uniflex 300:
Light liquid
Viscosity cSt @ 25C: 3300
Plasticizer for general purpose film, gaskets and tubing

310:
Light liquid
Viscosity cSt @ 25C: 5800
Plasticizer for high quality vinyl sheeting and rubber

312:
Light liquid
Viscosity cSt @ 25C: 980
Plasticizer for electrical tape and wiring, and other low temperature applications

314:
Light liquid
Viscosity cSt @ 25C: 5000
Medium MW plasticizer for use in PVA and nitrocellulose systems

315:
Light liquid
Viscosity cSt @ 25C: 6500
Medium MW plasticizer for heavy duty vinyl upholstery and refrigerator gaskets (low odor)

315X:
Light liquid
Viscosity cSt @ 25C: 9000
Higher MW plasticizer for wire and cable applications where extreme permmanence is needed

330:
Light liquid
Viscosity cSt @ 25C: 5300
Medium MW plasticizer for PVC applications where low odor, low taste and high permanence is needed

333:
Light liquid
Viscosity cSt @ 25C: 5700
Medium MW plasticizer for high quality vinyl sheeting

BASF Corp.: PALATINOL Plasticizers:

Palatinol 7P Phthalate:
   Formula: C22H34O4
   Molecular Weight: 362
   CAS registry number: 68515-44-6
   Palatinol 7P plasticizer is a linear phthalate ester based
on a predominantly linear C7 alcohol. Palatinol 7P can be used
as a primary plasticizer in flexible vinyl compounding or
added with other phthalate plasticizers to enhance fast fusion
applications. Palatinol 7P shortens both gelation and fusion
times while enhancing plastisol viscosity when compared to
equivalent branched alcohol chain phthalate plasticizers.
Product Specifications:
   Specific Gravity @ 25/25C: 0.986-0.994
   Ester content, by weight (% minimum): 99.0
   Acid Number, mg KOH/gm (max): 0.07
   Water, by weight (% max): 0.1
   Color, APHA (maximum): 25
   Clear oily liquid substantially free of foreign material

Palatinol 9P Di nonyl Phthalate:
   Formula: C26H42O4
   Molecular Weight: 418
   CAS Registry number: 68515-45-7
   Palatinol 9P plasticizer is a linear phthalate ester based
on a predominantly linear C9 alcohol. Palatinol 9P is designed
to meet today's more demanding requirements of vinyl in automotive and wire and cable industries.
   Palatinol 9P is an ideal plasticizer for use in automotive
interior trim and electrical applications. Because of its low
volatility, flexible vinyl products made with Palatinol 9P
exhibit excellent fogging characterictics. Its linearity offers
improved low temperature properties and ease of processibility
when compared to branched phthalate plasticizers.
Product Specifications:
   Specific Gravity @ 25/25C: 0.964-0.970
   Ester content, by weight (% min): 99.6
   Acid Number, mg KOH/gm (max): 0.07
   Water, by weight (% max): 0.1
   Color, APHA (max): 15
   Clear oily liquid substantially free of foreign material

BASF Corp.: PALATINOL Plasticizers (Continued):

Palatinol 11-9P-I High Molecular Weight Linear Phthalate:
  Formula: C28H46O4
  Molecular Weight: 458
  CAS Registry Numbers: 85507-79-5/111381-91-0/68515-45-7
  Palatinol 11-9P-I plasticizer is a linear phthalate ester
based on predominantly linear C9 and C11 alcohols. It exhibits
extremely low volatility in vinyl for automotive applications
that require low weight loss, long-term property retention and
excellent low temperature performance properties.
  Palatinol 11-9P-I is inhibited.
  Palatinol 11-9P-1 exhibits superior performance properties
over other 9-11 phthalates in vinyl. It is an excellent primary
plasticizer for:
  * Automotive interior trim parts
  * Instrument panel skins
  * Automotive wire harnesses
Product Specifications:
  Specific Gravity @ 25/25C: 0.950-0.958
  Ester Content, by weight (% min): 99.6
  Acid Number, mg KOH/gm (max): 0.07
  Water, by weight (% max): 0.1
  Color, APHA (max): 50
  Clear oily liquid substantially free of foreign material

Palatinol 11P-E (DUP):
  Formula: C30H50O4
  Molecular Weight: 474
  CAS Registry Number: 85507-79-5
  Palatinol 11P-E plasticizer is a linear phthalate ester
based on a predominantly linear C11 alcohol.
  Palatinol 11P-E exhibits very low volatility and good processing characteristics with vinyl resin. Its viscosity is relatively
low in relation to its molecular weight.
  Palatinol 11P-E provides the best low temperature performance
in vinyl compounds of any phthalate plasticizer.
  Palatinol 11P-E is inhibited.
  Vinyl resin compounders choose Palatinol 11P-E over DTDP for
its lower viscosity, greater resin compatibility, faster dry
blending and superior efficiency. When formulated to equal
concentrations, Palatinol 11P-E provides equal electrical
properties in 90C and 105C wire insulation compounds.
  With its lower viscosity, high thermal resistance, permanence
and low volatility, Palatinol 11P-E is well suited for demanding
applications such as:
  * Automotive interior (instrument panel skins)
  * Building wire insulation, 90C THW and 105C THW
  * Automotive wire harnesses
  * Vinyl roofing membranes
Product Specifications:
  Specific Gravity @ 25/25C: 0.948-0.955
  Ester content, by weight (% min): 99.6
  Acid Number, mg KOH/gm (max): 0.1

BASF Corp.: PALATINOL Plasticizers (Continued):

Palatinol 79P Phthalate Plasticizer:
    Formula: C24H38O4
    Molecular Weight: 398
    CAS Registry Numbers: 68515-45-7/68515-44-6/11381-89-6
    Palatinol 79P is a predominantly linear phthalate plasticizer
based upon C7 and C9 alcohols. The ratio of these alcohols and
their degree of linearity provide excellent processing characteristics, superior plastisol rheology and improved product
flexibility under low temperature conditions when compared
to DOP, DINP and DOTP in vinyl resins.
    Palatinol 79P is a general purpose phthalate plasticizer
and can replace DOP, DOTP and DINP in most applications as
well as replace general purpose blends that contain these
plasticizers. Palatinol 79P exhibits better efficiency and
imparts better low temperature flexibility than the above
mentioned plasticizers.
    It is particularly useful in plastisol applications because
it imparts lower initial viscosity, and better viscosity
stability than branched plasticizers.
    Typical end uses are:
    * Profile extrusions   * Shoe compounds
    * Traffic cones   * Fishing lures
    * Automotive sealants   * Fabric coatings
Product Specifications:
    Specific Gravity @ 25/25C: 0.971-0.981
    Ester Content, by weight (% min): 99.0

Palatinol 79TM-I Tri (heptyl, nonyl) Trimellitate:
    Formula: C33H54O6
    Molecular Weight: 547
    CAS Registry Number: 68515-60-6
    Palatinol 79TM-I is a high molecular weight trimellitate
ester based upon C7 and C9 predominantly linear alcohols.
The linearity of these alcohols imparts improved resistance
to volatility loss and better low temperature performance
than branch-chain trimellitates such as tri-2-ethylhexyl
trimellitate (TOTM).
    Palatinol 79TM-I is inhibited. The retention of physical
properties of the compounds pass Underwriters Laboratories
requirements for 105C rated cable.
    The excellent combination of permanence, low fogging, low
temperature performance and processibility, makes Palatinol
79TM-I an ideal plasticizer for vinyl electrical insulation
and automotive interior trim.
Product Specifications:
    Specific Gravity @ 25/25C: 0.982-0.987
    Ester Content, by weight (% Min): 99.0
    Acid Number, mg KOH/gm (Max): 0.2
    Water, by weight (% max): 0.1
    Color, APHA (max): 100
    Clear oily liquid substantially free of foregn material

BASF Corp.: PALATINOL Plasticizers (Continued):

Palatinol 711P Phthalate Plasticizer:
   Formula: C26H42O4
   Molecular Weight: 418
   Palatinol 711P is a predominantly linear phthalate plasticizer based upon C7, C9, and C11 alcohols. It is compatible with both homopolymer and copolymer vinyl resins, chlorinated rubber, SBR, neoprene and nitrile rubber as well as cellulosics. Palatinol 711P is used primarily to plasticize vinyl resin where good processing characteristics are needed and the finished product requires improved low temperature flexibility, low volatility or good outdoor weatherability.
   Palatinol 711P approved for medical applications.
   Palatinol 711P is an excellent general purpose phthalate plasticizer. With its efficiency, low-volatility, superior low-temperature flexibility and good weatherability, it is designed to offer the best balance of properties for both the vinyl processor and consumer.
   Typical end uses are:
   * Swimming pool and pond liners
   * Roofing membranes   * Coated fabrics
   * Tarpaulins   * High-end luggage
   * Exterior automotive trim
   * Building and communication wire
   Specific Gravity @ 25/25C: 0.965-0.973
   Ester Content, by weight (% min): 99.6

Palatinol N Diisononyl Phthalate:
   Formula: C26H42O4
   Molecular Weight: 418
   CAS Registry Number: 28553-12-0
   Palatinol N is an excellent primary plasticizer for plastic-ized vinyl articles. Its compatibility with vinyl is good even in high concentrations. Palatinol N in vinyl compounds provides good low temperature and low volatility performance. It has excellent rheological properties in plastisols and offers im-proved dry blend times.
   The mechanical properties of vinyl compounds plasticized with Palatinol N are comparable to products plasticized with other diisononyl phthalates. The unique nature of this ester results in lower volatility, improved processibility and better low temperature performance.
   The excellent resistance to water and outdoor exposure by vinyl plasticized with Palatinol N is an advantage in the manufacture of articles for the housing and construction markets.
   By virtue to its contribution to low viscosity and good visc-osity stability, Palatinol N is suitable for plastisols includ-ing spray coating, dipping, casting, or slush molding.
   Specific Gravity @ 25/25C: 0.970-0.976
   Ester Content, by weight (% min): 99.6

BASF Corp.: PALATINOL/PLASTOMOLL Plasticizers (Continued):

Palatinol TOTM Tri octyl Trimellitate:
Formula: $C33H54O6$
Molecular Weight: 547
CAS registry number: 3319-31-1
Palatinol TOTM is a primary branched monomeric plasticizer for vinyl homopolymer and copolymer resins. Palatinol TOTM is suggested for use in those end-use areas where extremely low volatility is required.
Palatinol TOTM can be blended with Palatinol 11P-E to optimize cost-performance in medium to high temperature compounds.
Palatinol TOTM provides desirable properties in vinyl applications which require good plasticizer/resin compatibility, low volatility, resistance to extraction by soapy water and good electrical properties.
Palatinol TOTM is often a good substitute for polyester polymeric plasticizers in low volatility applications where improvements in processing are desired.
Palatinol TOTM is suitable alone or in combination with Palatinol 11P-E for:
* 90C and 105C wire insulation
* Interior automotive applications (instrument panel skins)
Specific Gravity @ 25/25C: 0.984-0.991
Ester Content, by weight (% min): 99.0

Plastomoll GA Adipate Plasticizer:
Formula: $C24H46O4$
Molecular Weight: 419
CAS registry number: 215734-10-4
Plastomoll GA is an adipate plasticizer designed for use at the extremes of temperature performance. It combines exceptionally low volatility with good flexibility in vinyl film, sheet and plastisols at extremely low temperatures. This plasticizer may also be incorporated into rubber compounds.
Plastomoll GA can be used in plastisols to significantly reduce the coating viscosity. This will allow for a more uniform coating thickness and an improvement in coating line speed.
This linear-type diester is superior to DOA in low temperature properties and exhibits extremely low volatility for this class of plasticizer. It is suggested for use in products that have end use requirements demanding maximum flexibility at low temperature such as primary insulation meeting -40C CSA specifications and Type II bookbinding stock.
Plastomoll GA improves the low temperature flexibility of rubber. Its lower volatility provides better heat aging resistance than DOA in this application.
Specific Gravity @ 25/25C: 0.914-0.920
Ester content, by weight (% min): 99.6

ExxonMobil Chemical Co.: JAYFLEX Plasticizers:

Jayflex DHP (Dihexyl Phthalate):
   Distillation Range, C: Mid-boiling point @ 5 mmHg: 210
   Specific Gravity 20/20C: 1.007
   Viscosity @ 20C, cSt: 37
   Vapor Pressure @ 200C, mmHg: 3.0
   Density @ 20C, lb/gal: 8.38
   Flash Point, PMCC, F: 380
   CAS Registry Number: 68515-50-4

Jayflex 77 (Diisoheptyl Phthalate):
   Distillation Range, C: Mid-boiling point @ 5 mmHg: 220
   Specific Gravity 20/20C: 0.994
   Viscosity @ 20C, cSt: 51
   Vapor Pressure @ 200C, mmHg: 2.2
   Density @ 20C: lb/gal: 8.27
   Flash Point, PMCC, F: 390
   CAS Registry Number: 71888-89-6

Jayflex DIOP (Diisoctyl Phthalate):
   Distillation Range, C: Mid-boiling point @ 5 mmHg: 230
   Specific Gravity 20/20C: 0.985
   Viscosity @ 20C, cSt: 85
   Vapor Pressure @ 200C, mmHg: 1.0
   Density @ 20C, lb/gal: 8.20
   Flash Point, PMCC, F: 400
   CAS Registry Number: 27554-26-3

Jayflex DINP (Diisononyl Phthalate):
   Distillation Range, C: Mid-boiling point @ 5 mmHg: 245
   Specific Gravity 20/20C: 0.973
   Viscosity @ 20C, cSt: 102
   Vapor Pressure @ 200C, mmHg: 0.5
   Density @ 20C, lb/gal: 8.10
   Flash Point, PMCC, F: 415
   CAS Registry Number: 68515-48-0

Jayflex DIDP (Diisodecyl Phthalate):
   Distillation Range, C: Mid-Boiling Pont @ 5 mmHg: 256
   Specific Gravity 20/20C: 0.968
   Viscosity @ 20C, cSt: 129
   Vapor Pressure @ 200C, mmHg: 0.35
   Density @ 20C, lb/gal: 8.06
   Flash Point, PMCC, F: 435
   CAS Registry Number: 68515-49-1

ExxonMobil Chemical Co.: JAYFLEX Plasticizers (Continued):

Jayflex UDP (Undecyl, Dodecyl Phthalate):
    Distillation range, C: Mid-boiling point @ 5mmHg: 275
    Specific gravity 20/20C: 0.959
    Viscosity @ 20C, cSt: 265
    Vapor Pressure @ 200C, mmHg: 0.15
    Density @ 20C, lb/gal: 7.98
    Flash Point PMCC, F: 440
    CAS registry number: 68515-47-9

Jayflex DTDP (Ditridecyl Phthalate):
    Distillation range, C: Mid-boiling Point @ 5 mmHg: 286
    Specific gravity 20/20C: 0.957
    Viscosity @ 20C, cSt: 322
    Vapor Pressure @ 200C, mmHg: 0.08
    Density @ 20C, lb/gal: 7.96
    Flash point PMCC, F: 445
    CAS registry number: 68515-47-9

Jayflex L9P (Di-L-Nonyl Phthalate):
    Distilation range, C: Mid-boiling point @ 5 mmHg: 250
    Specific gravity 20/20C: 0.969
    Viscosity @ 20C, cSt: 54
    Vapor Pressure @ 200C, mmHg: 0.5
    Density @ 20C, lb/gal: 8.06
    Flash Point PMCC, F: 415
    CAS registry number: Mixture

Jayflex L911P (Di-L-(Nonyl, Undecyl) Phthalate):
    Distillation range, C: Mid-boiling point @ 5 mmHg: 261
    Specific gravity 20/20C: 0.962
    Viscosity @ 20C, cSt: 74
    Vapor Pressure @ 200C, mmHg: <0.3
    Density @ 20C, lb/gal: 8.01
    Flash Point PMCC, F: 419
    CAS registry number: Mixture

Jayflex L11P (Di-L-Undecyl Phthalate):
    Distillation range, C: Mid-boiling point @ 5 mmHg: 271
    Specific gravity 20/20C: 0.954
    Viscosity @ 20C, cSt: 77
    Vapor Pressure @ 200C, mmHg: <0.15
    Density @ 20C, lb/gal: 7.94
    Flash Point PMCC, F: 460
    CAS registry number: 3648-20-2

Jayflex DIOA (Diisooctyl Adipate):
    Specific gravity 20/20C: 0.928
    Density @ 20C, lb/gal: 7.73
    Flash point PMCC, F: 405
    CAS Registry Number: 1330-86-5

ExxonMobil Chemical Co.: JAYFLEX Plasticizers (Continued):

Jayflex DINA (Diisononyl Adipate):
   Distillation Range, C: Mid-boiling point @ 5 mmHg: 233
   Specific gravity 20/20C: 0.922
   Viscosity @ 20C, cSt: 22
   Vapor pressure @ 200C, mmHg: <2.0
   Density @ 20C, lb/gal: 7.67
   Flash Point, PMCC, F: 390
   CAS Registry Number: 33703-08-1

Jayflex DIDA (Diisodecyl Adipate):
   Specific gravity 20/20C: 0.918
   Density @ 20C, lb/gal: 7.65
   Flash Point, PMCC, F: 437
   CAS Registry Number: 27178-16-1

Jayflex DTDA (Ditridecyl Adipate):
   Specific gravity 20/20C: 0.913
   Density @ 20C, lb/gal: 7.61
   Flash Point, PMCC, F: 455
   CAS Registry Number: 16958-92-2

Jayflex TIOTM (Triisooctyl Trimellitate):
   Distillation Range, C: Mid-boiling point @ 5 mmHg: 300
   Specific gravity 20/20C: 0.992
   Viscosity @ 20C, cSt: 312
   Density @ 20C, lb/gal: 8.26
   Flash Point, PMCC, F: 430
   CAS Registry Number: 27251-75-8

Jayflex TINTM (Triisononyl Trimellitate):
   Distillation Range, C: Mid-boiling point @ 5 mmHg: 311
   Specific gravity 20/20C: 0.978
   Viscosity @ 20C, cSt: 430
   Density @ 20C, lb/gal: 8.14
   Flash Point, PMCC, F: 465
   CAS Registry Number: 53894-23-8

Jayflex 210 (Naphthenic Hydrocarbon):
   Distillation Range, C: Mid-boiling point @ 5 mmHg: 330
   Specific gravity 20/20C: 0.870
   Viscosity @ 20C, cSt: 28
   Density @ 20C, lb/gal: 7.24
   Flash Point, PMCC, F: 295
   CAS Registry Number: 64742-53-6

Jayflex 215 (Aliphatic Hydrocarbon):
   Distillation Range, C: Mid-boiling Point @ 5 mmHg: 271
   Specific gravity 20/20C: 0.762
   Viscosity @ 20C, cSt: 12
   Density @ 20C, lb/gal: 6.34
   Flash Point, PMCC, F: 250
   CAS Registry Number: 64771-72-8

Harwick Standard Distribution Corp.: Plasticizers:

Adipates:
Polycizer DOA:
Merrol DOA:
   Di-2 ethylhexyl adipate
   Polymer Usage: R-1,2/P-1,2
   General Purpose, Low Temperature/Flexibility
   Miscellaneous: FDA, low water extraction, UV stability

Merrol DIDA:
   Diisodecyl adipate
   Polymer Usage: R-1,2/P-1,2
   Low Temperature/Flexibility, Low Volatility
   Miscellaneous: Good compatability

Merrol DTDA:
   Ditridecyl adipate
   Polymer Usage: P-1,2,3
   Low Volatility, Low Extraction

Merrol 4206 (DBEA):
   Dibutoxyethyl adipate
   Polymer Usage: R-1,2,3/P-2
   Low Temperature/Flexibility

Polycizer DBEEA:
Merrol 4226:
   Dibutoxyethoxyethyl adipate
   Polymer Usage: R-1,2,3
   Low Temperature/Flexibility, Low Volatility, Low Extraction,
Heat Aging Resistance

Merrol 4426:
   TEG monobutylether adipate
   Polymer Usage: R-1,2,3
   Low Temperature/Flexibility, Low Volatility, Low Extraction

Merrol 79A:
   Dialkyl adipate
   Polymer Usage: R-1,2/P-1
   Low Temperature/Flexibility

Azelates:
Merrol DOZ:
   Di-2-ethylhexyl azelate
   Polymer Usage: R-1,2/P-1,2
   General Purpose, Low Temperature/Flexibility, Low Volatility
   Miscellaneous: Excellent low temp.

Harwick Standard Distribution Corp.: Plasticizers (Continued):

Benzoates:
Benzoflex 9-88/Polycizer DP:
   Dipropylene glycol dibenzoate
   Polymer Usage: R-1/P-1,2
   Low Volatility, High Solvating
   Miscellaneous: Polyurethanes

Benzoflex 9-88 SG:
   Dipropylene glycol dibenzoate
   Polymer Usage: R-1/P-1,2
   High Solvating
   Miscellaneous: Castable polyurethanes, urethane sealants

Benzoflex 50/Polycizer DP 500:
   Diethylene/dipropylene glycol dibenzoate
   Polymer Usage: R-1/P-1,2
   Low Volatility, High Solvating
   Miscellaneous: Water-based adhesives

Benzoflex 2088:
   Diethylene glycol dibenzoate, triethylene glycol dibenzoate,
dipropylene glycol dibenzoate
   Polymer Usage: R-1/P-1,2
   Low Volatility, Low Extraction, High Solvating
   Miscellaneous: High solvator, low VOC's, FDA

Chlorinated Paraffins:
Plastichlor P-Series:
   Liquid chlorinated paraffins
   Polymer Usage: R-2/P-1
   General Purpose, Flame Resistance

Mono-Esters:
Polycizer Butyl Oleate/Merrol 418T:
   N-butyl oleate
   Polymer Usage: R-2/P-2
   Low Temperature/Flexibility
   Primary plasticizer for polychloroprene

Polycizer MO:
   Vegetable oil
   Polymer Usage: R-2
   Low Temperature/Flexibility, Low Volatility, Heat Aging
Resistance, High Solvating
   Miscellaneous: Low and high temp. for polychloroprene

Harwick Standard Distribution Corp.: Plasticizers (Continued):

Mono-Esters (Continued):
Polycizer OLN:
   Oleyl nitrile
   Polymer Usage: R-1
   Low Extraction, High Solvating

Natroflex IOT:
   Isocetyl tallate
   Polymer Usage: R-1,2
   General Purpose, Low Temperature Flexibility

Merrol 818T:
   Alkyl tallate
   Polymer Usage: R-1, P-2
   General Purpose, Low Temperature Flexibility

Petroleum Process Oils:
Stan-Lube Series:
   Paraffinic oils
   Polymer Usage: Non-polar
   General Purpose
   Miscellaneous: Light color, good for EPR's

Stan-Plas Series:
   Naphthenic oils
   Polymer Usage: R-1
   General Purpose

Enerflex 121:
   Aromatic oils
   Polymer Usage: R-1,2
   General Purpose, High Solvating
   Miscellaneous: Excellent for NR, BR, SBR, FDA

Duoprime Series:
   White oils
   Polymer Usage: Non-polar
   General Purpose
   Miscellaneous: FDA

Harwick Standard Distribution Corp.: Plasticizers (Continued):

Phosphate Esters:
   Used to partially replace process oils and plasticizers, and
aid flame retardants by forming polyphosphoric acid char which
inhibits flame propagation, reduces substrate temperature and
acts as an oxygen barrier.

Lindol/Merrol TCP:
   Tricresyl phosphate
   Polymer Usage: P-1,2
   General Purpose, Low Volatility, Flame Resistance

Phosflex 4:
   Tributyl phosphate
   Polymer Usage: R-2/P-1,2
   Flame Resistance, High Solvating

Phosflex 41-P/Merrol 521:
   Isopropylated triaryl phosphate
   Polymer Usage: R-1,2/P-1
   Flame Resistance

Phosflex T-BEP/Merrol T-BEP:
   Tributoxyethyl phosphate
   Polymer Usage: R-1,2,3/P-1,2
   Low Temperature/Flexibility, Flame Resistance, High Solvating

Phosflex 71 B:
   Butylated triphenyl phosphate
   Polymer Usage: R-1,2/P-1
   Flame Resistance

Phosflex 362:
   2-ethylhexyl diphenyl phosphate
   Polymer Usage: R-1,2/P-1,2
   Flame Resistance

Phosflex 370:
   Blend isodecyl diphenyl phosphate & t-butylated triphenyl
phosphate
   Polymer Usage: R-1,2/P-1
   Flame Resistance

Phosflex 390:
   Isodecyl diphenyl phosphate
   Polymer Usage: R-1,2/P-1,2
   Flame Resistance

Merrol TOF:
   Trioctyl phosphate
   Polymer Usage: P-1
   Flame Resistance

Harwick Standard Distribution Corp.: Plasticizers (Continued):

Phthalates:
Merrol DAP:
   Diallyl phthalate
   Polymer Usage: R-1,2/P-3
   High Solvating
   Miscellaneous: Co-curing

Polycizer DBP/Merrol DBP:
   Di-n-butyl phthalate
   Polymer Usage: R-1,2/P-1,2
   General Purpose, High Solvating
   Miscellaneous: Good emollient for cosmetics

Polycizer DIDP/Merrol DIDP:
   Diisodecyl phthalate
   Polymer Usage: R-1,2/P-1,2
   Low Volatility, Low Extraction
   Miscellaneous: Also E grade

Polycizer DINP/Merrol DINP:
   Diisononyl phthalate
   Polymer Usage: R-1,2/P-1,2
   Low Volatility

Polycizer DOP/Merrol DOP:
   Di-2-ethylhexyl phthalate
   Polymer Usage: R-1,2/P-1,2
   General Purpose

Polycizer DUP:
   Diiundecyl phthalate
   Polymer Usage: R-1,2/P-1,2
   Low Temperature/Flexibility, Low Volatility, Heat Aging
Resistance
   Miscellaneous: Low fogging

Polycizer DOTP:
   Di-2-ethylhexyl terephthalate
   Polymer Usage: R-1,2/P-1,2
   General Purpose, Low Temperature/Flexibility
   Miscellaneous: Also E grade

Harwick Standard Distribution Corp.: Plasticizers (Continued):

Polymerics:
Admex P-27:
   Polyester adipate
   Polymer Usage: R-1/P-1,2
   Migration Resistance
   Miscellaneous: High purity, good electrical properties

Admex 409:
   Polyester adipate
   Polymer Usage: R-1/P-1,2
   General Purpose, Migration Resistance, Heat Aging
   Miscellaneous: Good electrical properties

Admex 412:
   Polyester adipate
   Polymer Usage: R-1/P-1
   Low Temperature/Flexibility, Permeability
   Miscellaneous: Low viscosity, easy processing

Admex 429:
   Polyester adipate
   Polymer Usage: R-1/P-1,2
   Migration Resistance
   Miscellaneous: Non-fogging, humidity resistance

Admex 523:
   Mixed polyester
   Polymer Usage: R-1/P-1,2
   General Purpose, Migration Resistance, Low Extraction
   Miscellaneous: Low viscosity

Admex 760:
   Polyester adipate
   Polymer Usage: R-1,2/P-1,2
   Permeability, Migration Resistance
   Miscellaneous: Excellent permanence, low water extractability

Admex 761:
   Polyester adipate
   Polymer Usage: R-1/P-1,2
   Low Extraction

Admex 770:
   Mixed polyester
   Polymer Usage: R-1,2/P-1,2
   Permeability, Migration Resistance
   Miscellaneous: Excellent weatherability (decals)

Harwick Standard Distribution Corp.: Plasticizers (Continued):

Polymerics (Continued):
Admex 775:
   Mixed polyester
   Polymer Usage: R-1/P-1,2

Admex 910-001:
   Mixed polyester
   Polymer Usage: R-1/P-1,2
   Low Extraction
   Miscellaneous: Low water extraction

Admex 1723:
   Mixed polyester
   Polymer Usage: R-1/P-1,2
   Permeability
   Miscellaneous: Printability

Admex 2632:
   Mixed polyester
   Polymer Usage: R-1/P-1,2
   General Purpose
   Miscellaneous: FDA

Admex 6187:
   Polyester adipate
   Polymer Usage: R-1/P-1,2
   Migration Resistance, Low Extraction
   Miscelleneous: Solvent & oil resistance

Admex 6985:
   Polyester adipate
   Polymer Usage: R-1/P-1,2
   Migration Resistance, Low Extraction, Heat Aging
   Miscellaneous: Very low volatility

Admex 6994:
   Mixed polyester
   Polymer Usage: R-1/P-1,2
   Migration Resistance
   Miscellaneous: Mar resistance, low fogging

Admex 6995:
   Polyester adipate
   Polymer Usage: R-1/P-1,2
   Permeability
   Miscellaneous: UV weatherability

Admex 6996:
   Polyester adipate
   Polymer Usage: R-1/P-1,2
   Low Temperature/Flexibility
   Miscellaneous: Printability

Harwick Standard Distribution Corp.: Plasticizers (Continued):

Polymerics (Continued):
Merrol P-1030 LV:
    Polyester sebacate
    Polymer Usage: R-1/P-1,2
    Low Extraction, Heat Aging Resistance
    Miscellaneous: Low viscosity, controls oil swell

Merrol P-5511:
    Polyester nylonate
    Polymer Usage: R-1/P-1
    Low Extraction

Merrol P-6303:
    Polyester adipate
    Polymer Usage: R-1,2,3/P-1,2
    Low Extraction
    Miscellaneous: High viscosity, low migration

Polycizer PE-320/Merrol P-6310:
    Polyester adipate
    Polymer Usage: R-1,2,3/P-1
    General Purpose
    Miscellaneous: Good balance of properties, easy to mix

Polycizer PE-312/Merrol P-6320:
    Polyester adipate
    Polymer Usage: R-1,2/P-1
    Low Temperature/Flexibility, Low Extraction
    Miscellaneous: Solvent & oil resistance, low temp flexibility

Merrol P-6412:
    Polyester adipate
    Polymer Usage: R-1,2/P-1,2
    Low Extraction
    Miscellaneous: Medium viscosity, FDA

Polycizer PE-330/Merrol P-6410:
    Polyester adipate
    Polymer Usage: P-1,2
    Low Volatility, Low Extraction
    Miscelleneous: Polystyrene & ABS

Merrol P-6420:
    Polyester adipate
    Polymer Usage: P-1
    Low Extraction
    Miscellaneous: Good color

Harwick Standard Distribution Corp.: Plasticizers (Continued):

Polymerics (Continued):
Merrol P-6422:
   Polyester adipate
   Polymer Usage: P-1
   General Purpose, Low Temperature/Flexibility, Low Volatility
   Miscellaneous: Low viscosity, plastisols

Merrol P-6424:
   Polyester adipate
   Polymer Usage: R-1,2,3/P-1
   Low Extraction, Heat Aging Resistance, High Solvating

Merrol P-8227:
   Mixed polyester
   Polymer Usage: R-1,2/P-1,2
   General Purpose, Low Temperature/Flexibility, Heat Aging
Resistance
   Misc.: Excellent high temp, wire & cable

Merrol P-8425-C:
   Polyester phthalate
   Polymer Usage: R-1,2/P-1,2
   Low Temperature/Flexibility, Low Volatility, Heat Aging
Resistance
   Misc.: Excellent humid age resistance

Sebacates:
Polycizer DBS/Merrol DBS:
   Di-n-butyl sebacate
   Polymer Usage: R-1,2/P-1,2
   Low Temperature/Flexibility, High Solvating
   Miscellaneous: FDA

Polycizer DOS/Merrol DOS:
   Di-2-ethylhexyl sebacate
   Polymer Usage: R-2/P-1,2
   General Purpose, Low Temperature/Flexibility, Low Extraction
   Miscellaneous: Low temp greases & caulks

Merrol 4200 (DBES):
   Dibutoxyethyl sebacate
   Polymer Usage: R-1,2/P-1,2
   Low Temperature/Flexibility, Low Volatility
   Miscellaneous: Combines low-temp & low volatility

Harwick Standard Distribution Corp.: Plasticizers (Continued):

Specialty:
Merrol N-302:
   N-ethyl ortho/para-toluenesulfonamide
   Polymer Usage: R-3/P-1,2,3
   Migration Resistance
   Miscellaneous: Adhesion

Merrol N-303:
   N-butylbenzene sulfonamide
   Polymer Usage: R-3/P-1,2,3
   General Purpose

Plasticizer SC-B/Merrol 3810:
   Triethyleneglycol dicaprate/caprylate
   Polymer Usage: R-1,2,3
   Low Temperature/Flexibility
   Miscellaneous: FDA

Plasticizer SC-E:
   Triethyleneglycol di-2-ethylhexoate
   Polymer Usage: R-1,2,3
   Low Temperature/Flexibility
   Miscellaneous: Flexibility over a wide temp range

Merrol 4221 (DBEEF):
   Dibutoxyethoxyethyl formal
   Polymer Usage: R-1,2,3
   Low Temperature/Flexibility

Polycizer TXIB:
   2,2,4-trimethyl-1,3-pentanediol diisobutyrate
   Polymer Usage: P-1,2
   General Purpose, Heat Aging Resistance
   Miscellaneous: Excellent PVC processing

Hercoflex 600:
   Pentaerythritol ester of fatty acids
   Polymer Usage: R-1,2
   Low Temperature/Flexibility, Permeability, Migration
Resistance, Low Extraction, Heat Aging Resistance
   Miscelleneous: Excellent low and high temp

Hercoflex 707/707A:
   Pentaerythritol ester of fatty acids
   Polymer Usage: R-1,2
   Low Temperature/Flexibility, Permeability, Migration
Resistance, Low Extraction, Heat Aging Resistance
   Miscellaneous: Excellent low and high temp

Harwick Standard Distribution Corp.: Plasticizers (Continued):

Specialty (Continued):
Polycizer TOTM/Merrol TOTM:
   Tri-2-ethylhexyl trimellitate
   Polymer Usage: R-1,2/P-1,2
   Permeability, Low Extraction, Heat Aging Resistance
   Miscellaneous: Also E&CA grades, excellent water resistance

Merrol 810TM:
   Tri(n-octyl/n-decyl) trimellitate
   Polymer Usage: R-2
   Low Temperature/Flexibility, Permeability, Low Extraction,
Heat Aging Resistance
   Miscellaneous: Oxidation resistance, excellent water resistance

Polycizer ESO/Merrol E-68:
   Epoxidized soybean oil
   Polymer Usage: R-1/P-1,2,3
   Permeability, Migration Resistance, Heat Aging Resistance
   Miscellaneous: Good heat stabilizer

Polymer Usage Key

R-1: NBR, NBR/PVC	P-1: PVC
R-2: CR, CPE, CSM	P-2: PVAC, PS, ABS, Cellulosics
R-3: ECO, Fluoroelastomers, Polyacrylates	P-3: Eng. Resins, Polyester, Alloys

Inspec UK Ltd.: BISOFLEX Plasticizers:

Bisoflex TOPM:
  Specialty High Temperature Plasticizer
  A very high performance alkyl pyromellitate ester intended
for use in PVC and synthetic cable formulations where excell-
ent resistance to heat aging is a requirement. It is also
intended for use in flexible PVC medical items taking advantage
of its high degree of compatibility and extraction resistance.
  Chemically Bisoflex TOPM is tetra-2-ethylhexyl pyromellitate
ester, with Chemical Abstract Service Number 3126-80-5.
Typical Specification:
  Appearance: Clear liquid free from suspended solids
  Colour: 250 Max: Pt/Co
  Saponification Number: 317 mg KOH/g
  Acidity: 0.005 % Mass
  Density @ 20C: 0.985 Kg/litre
  Water Content: 0.02 max. % mass
Physical Properties:
  Molecular mass: 702
  Density at 20C: Kg/litre: 0.985
  Flash point, COC: C: 271
  Pour Point: C: -48
  Viscosity at 40C: cSt: 172
  Viscosity at 100C: cSt: 16

Bisoflex TOT (MED):
  Bisoflex TOT (med) is primarily intended for use in flexible
PVC medical items such as blood bags and dialysis tubing.
Bisoflex TOT (med) is characterized by its excellent biocompa-
tibility and low in vivo extraction.
  Chemically, Bisoflex TOT (med) is a high purity low odour
trimellitate of 2-ethylhexanol. It contains no antioxidants
which are typically used in technical grade trimellitates.
  Its Chemical Astracts Service Number is 3319-31-1.
Typical Specification:
  Appearance: Clear and solids free
  Colour: PtCo: 75 max
  Density at 20C: kg/litre: 0.984-0.990
  Ester content: % mass: 97.5 min
  Acidity: % mass: 0.02 max
  Water content: % mass: 0.1 max
  Residual alcohol: % mass: 0.1 max
Typical Physical Properties:
  Molecular Mass: 547
  Density at 20C: kg/litre: 0.987
  Flash point, Pensky Martens Open Cup: C: 271
  Pour Point, ASTM: C: -30
  Absolute Viscosity at 20C: mPa.s: 300
  Refractive index at 20C: 1.485

Inspec UK Ltd.: BISOFLEX Plasticizers (Continued):

Bisoflex 102:
An excellent low temperature plasticizer for a wide range
of polymers, particularly nitrile rubber, chloroprene rubber
and polyvinyl chloride. It is essentially the ester of triethylene glycol with linear acids, average carbon number C9 and
gives superior low temperature performance to the adipates and
the sebacates.
Bisoflex 102 also reduces the surface and volume resistivity
of polyvinyl chloride compositions. This is an important factor
in the manufacture of coal mine conveyor belting, hospital
flooring and footwear where the generation of static electricity
must be avoided.
Typical Specification:
  Appearance: Clear and free from matter in suspension
  Colour, Pt/Co: max: 60
  Density at 20C: kg/litre: 0.960-0.966
  Saponification number: mg.KOH/g min: 258-270
  Acidity (as C9 acid): % mass max: 0.09
  Water content: % mass max: 0.1
Properties: Physical Compounds:
  Molecular mass (approximate): 426
  Boiling point at 0.3mbar: C: 220
  Flashpoint Pensky Martens open cup: C: >200
  Pour Point ASTM (Crystallizing Point) C: -5
  Refractive index: 1.447
  Absolute viscosity at 20C: cP: 22

Bisoflex DOA:
An efficient low temperature plasticizer for vinyl chloride
polymers and copolymers and the more polar synthetic rubbers.
Its Chemical Abstract Service Number is 103-23-1.
Typical Specification:
  Appearance: Clear and free of suspended matter
  Colour, Pt/Co: max: 50
  Density at 20: kg/litre: 0.924-0.928
  Ester Content: % mass min: 99.0
  Acidity (as adipic acid): % mass: 0.025
  Water Content: % mass max: 0.1
Physical Properties:
  Molecular mass: 371
  Density at 20C: kg/litre: 0.926
  Boiling Point at 13.3 mbar: C: 232
  Vapour Pressure at 200C: mbar: 3.3
  Flash point, Pensky Martens open cup: C: 213
  Pour Point: C: <-60
  Refractive Index: 1.447
  Absolute Viscosity at 20C: cP: 13

Inspec UK Ltd: BISOFLEX Plasticizers (Continued):

Bisoflex 120:
    An efficient low temperature plasticizer for PVC and its
copolymers and the more polar synthetic rubbers. Bisoflex 120
is an economical replacement for dioctyl azelate (DOZ) and
dioctyl sebacate (DOS). These low temperature plasticizers,
based on natural product-derived acids are frequently used for
applications where dioctyl adipate (DOA) is considered too
volatile. The plasticizing, processing and volatility characteristics of Bisoflex 120 are very similar to those of DOZ.
In PVC Bisoflex 120 would be used typically in conjunction
with a dialkyl phthalate general purpose plasticizer. The major
application for Bisoflex 120 is as a plasticizer for PVC plastisols employed in steel coil coatings.
Typical Specification:
    Appearance: Clear and free of suspended matter
    Colour, Lovibond 152mm cell: 2.0Y 0.5R max
    Acidity as adipic acid: % w/w: 0.02 max
    Specification Number: mgKOH/g: 260-275
    Water content: % max: 0.1 max
    Density at 20C: kg/litre: 0.915-0.930
Physical Properties:
    Refractive Index: 1.450
    Viscosity at 20C: cP: 23
    Flash point, Pensky Martens open cup: C: 230
    Pour Point: C: -26

Bisoflex 124:
    An excellent antistatic plasticizer for PVC and synthetic
rubbers. It avoids the thermal stability problems associated
with ionic antistatic additives. In addition, Bisoflex 124
confers excellent low temperature properties and low
plastisol viscosity.
    Bisoflex 124 has considerably higher compatibility with PVC
then conventional non-ionic surfactant type antistatic agents.
    The components of Bisoflex 124 are listed in the European
Inventory of Existing Commercial Chemical Substances (EINECS)
and the US Toxic Substances Control Act (TSCA) Chemical
Substances Inventory.
Typical Specification:
    Appearance: Amber liquid, solids free
    Density at 20C: kg/litre: 1.015-1.025
    Acid number: mg.KOH/g: 5 max
    Saponification number: mg.KOH/g: 205-225
    Water content: %: 0.5 max
Physical Properties:
    Flash point, Pensky Martens closed cup: C: 150
    Freezing point: C: <-10
    Refractive Index: 1.4550
    Absolute viscosity at 20C: cP: 32

Inspec UK Ltd.: BISOFLEX Plasticizers (Continued):

Bisoflex DBS:
   DBS is an excellent low temperature plasticizer for synthetic rubbers and vinyl chloride polymers and copolymers. It is also an effective process plasticizing aid for polyvinylidene chloride (PVdC) copolymers.
   Chemically DBS is di-butyl sebacate. Its Chemical Abstract Service Number is 109-43-3.
Typical Specification:
   Appearance: Clear liquid free of visible particles
   Colour: Pt/Co: 50 max.
   Density @ 20C: kg/litre: 0.934-0.940
   Total Ester Content: % mass: 99.0 min.
   Acidity (as sebacic acid): % mass: 0.02 max.
   Water Content: % mass: 0.10 max.
Typical Physical Properties:
   Molecular weight: 314
   Viscosity @ 25C: mPa.s: 12-16
   Flash Point (PMCC): C: 180
   Pour Point: C: -11
   Refractive index at 20C: 1.44

Bisoflex DOS:
   DOS is a low temperature plasticizer and lubricant base stock.
   Chemically DOS is di-2-ethylhexyl sebacate.
   Its Chemical Abstract Service Number is 122-62-3
Typical Specification:
   Appearance: Clear liquid free of visible particles
   Colour: 50 max.: Pt/Co
   Acidity: 0.01 max.: % mass
   Ester Content: 97.5 min.: % mass
   Water: 0.05 max.: % mass
   Density at 20C: 0.912-0.917 kg/litre
Physical Properties:
   Molecular Weight: 426
   Viscosity at 20C: mPas: 18.5-21
   Pour Point: C: -67
   Flash Point (PMOC): C: >200
   Refractive Index at 20C: 1.4510

Inspec UK Ltd.: BISOFLEX Plasticizers (Continued):

Bisoflex DL79A:
  Bisoflex DL79A is a very efficient low temperature plasticizer for PVC and its copolymers and the more polar synthetic rubbers. It confers slightly better low temperature properties and is less volatile than DOA (Di-2-ethylhexyl adipate).
  Its Chemical Abstract Service Number is 68515-75-3.
Typical Specification:
    Appearance: Clear and free of suspended matter
    Colour, Lovibond 6" cell: 1.5Y 0.5R max
    Acidity as adipic acid: % max: 0.02
    Saponification number: mg.KOH/g: 300-315
    Water content: % max.: 0.1 max
    Density at 20C: 0.92-0.93
Physical Properties:
    Molecular Mass: 360
    Density at 20C: 0.925
    Refractive Index: 1.447
    Viscosity at 20C: mPa.s: 16
    Viscosity at 100C: mPa.s: 3.0
    Flash Point, Pensky Martens open cup: 216C

CK Witco Corp.: DRAPEX Epoxy Plasticizers:

Drapex 4.4 Epoxy Plasticizer:

Drapex 4.4 epoxy plasticizer is an epoxidized octyl tallate which imparts low temperature flexibility as well as excellent heat and light stability to vinyl compounds.

In combination with certain vinyl stabilizers, such as the MARK barium/cadmium types, the use of Drapex 4.4 plasticizer results in a synergistic improvement of this stability at low cost.

Drapex 4.4 also provides vinyl plastisol and organosol formulations with low initial viscosity and extended viscosity stability by virtue of its low solvating action.

Properties: Typical Values:
    Color (Gardner): 1
    Specific Gravity @ 25C (77F): 0.920
    Oxirane Oxygen: 4.7%
    Iodine Value (HANUS): 2.5
    Refractive Index @ 25C (77F): 1.4570
    Acid Value: 0.5
    Moisture: 0.03%
    Freezing Point: -20C (-4F)
    Odor: Faintly Fatty
    Viscosity @ 25C (77F): Brookfield: 20 cps
    Flash Point (COC): 220C (428F)
    Water Solubility @ 20C (68F): 0.0% by wt.
    Molcular Weight (approx.): 420

Drapex 6.8 Epoxy Plasticizer:

Drapex 6.8 epoxy plasticizer is an epoxidized soybean oil specifically designed to provide optimum heat and light stabilizing performance together with maximum compatibility in vinyl compounds.

Properties: Typical Values:
    Physical State: Liquid
    Color (Gardner): 1
    Specific Gravity @ 25C (77F): 0.992
    Oxyrane Oxygen: 7.0%
    Iodine Value (HANUS): 1.6
    Refractive Index @ 25C (77F): 1.4710
    Acid Value: 0.5
    Moisture: 0.02%
    Viscosity @ 25C (77F) (Gardner): M
    Viscosity @ 25C (77F) (Brookfield): 320 cps
    Pour Point: 0C (32F)
    Flash Point (COC): 290C (554F)
    Odor: Faintly Fatty
    Water Solubility @ 20C (68F): 0.01% by wt
    Molecular Weight (approx.): 1000

CKWitco Corp.: DRAPEX Epoxy Plasticizers (Continued):

Drapex 10.4 Epoxy Plasticizer:
Drapex 10.4 plasticizer is designed specifically to provide non-toxic vinyl compounds with optimum heat stabilizing action when used with primary metallic stabilizers such as ARGUS' line of calcium-zinc additives. It provides one of the highest oxirane oxygen contents available.
Properties: Typical Values:
    Color (Gardner): 1
    Specific Gravity @ 25C (77F): 1.030
    Oxyrane Oxygen: 9.0%
    Iodine Value (Hanus): 2.5
    Refractive Index @ 25C (77F): 1.4765
    Acid Value: 0.65
    Moisture: 0.08%
    Viscosity @ 25C (Gardner): V-
    Viscosity @ 25C (Brookfield): 1000 cps
    Pour Point: -5C (23F)
    Flash Point: 290C (554F)
    Odor: Faintly Fatty
    Water Solubility @ 20C (68F): 0.1% by wt
    Molecular Weight (approx): 1000

Morflex, Inc.: CITROFLEX Citric Acid Esters:

Citroflex citric acid esters provide a wide range of benefits when used as plasticizers with aqueous- and solvent-based polymers, including acrylic, methacrylic, ethyl cellulose, hydroxypropyl methyl cellulose, nitrocellulose, vinyl acetate, vinyl chloride, vinyl pyrrolidone, vinylidene chloride, and urethane polymer systems.

Citroflex:
    C-2: Triethyl Citrate
    A-2: Acetyltriethyl Citrate
    C-4: Tri-n-butyl Citrate
    A-4: Acetyltri-n-butyl Citrate
    A-6: Acetyltri-n-hexyl Citrate
    B-6: n-Butyryltri-n-hexyl Citrate

Product Data:
C-2:
    CAS Number: 77-93-0
    Molecular Weight: 276.3
    Molecular Formula: $C_{12}H_{20}O_7$

A-2:
    CAS Nunber: 77-89-4
    Molecular Weight: 318.3
    Molecular Formula: $C_{14}H_{22}O_8$

C-4:
    CAS Number: 77-94-1
    Molecular Weight: 360.4
    Molecular Formula: $C_{18}H_{32}O_7$

A-4:
    CAS Number: 77-90-7
    Molecular Weight: 402.5
    Molecular Formula: $C_{20}H_{34}O_8$

A-6:
    CAS Number: 24817-92-3
    Molecular Weight: 486
    Molecular Formula: $C_{26}H_{46}O_8$

B-6:
    CAS Number: 82469-79-2
    Molecular Weight: 514
    Molecular Formula: $C_{28}H_{50}O_8$

Morflex, Inc.: CITROFLEX Citric Acid Esters (Continued):

C-2:
   Appearance: Clear Liq.
   Odor: Essentially odorless
   Pour Point, C: -45
   Viscosity @ 25C, cps: 35
   Volatiles, %: 1.6
   Vapor Pressure, mm Hg @ 20C: 6.4 x 10 -3
   Flash Point (COC), C: 155

A-2:
   Appearance: Clear Liq.
   Odor: Essentially odorless
   Pour Point, C: -43
   Viscosity @ 25C, cps: 54
   Volatiles, %: 1.3
   Vapor Pressure, mm Hg @ 20C: 5.7 x 10 -3
   Flash Point (COC), C: 188

C-4:
   Appearance: Clear Liq.
   Odor: Essentially odorless
   Pour Point, C: -62
   Viscosity @ 25C, cps: 32
   Volatiles, %: 0.2
   Vapor Pressure, mm Hg @ 20C: 9.6 x 10 -2
   Flash Point (COC), C: 185

A-4:
   Appearance: Clear Liq.
   Odor: Essentially odorless
   Pour Point, C: -59
   Viscosity @ 25C, cps: 33
   Volatiles, %: 0.2
   Vapor Pressure, mm Hg @ 20C: 5.2 x 10 -2
   Flash Point (COC), C: 204

A-6:
   Appearance: Clear Liq.
   Odor: Essentially odorless
   Pour Point, C: -57
   Viscosity @ 25C, cps: 36
   Volatiles, %: 1.4
   Flash Point (COC), C: 240

B-6:
   Appearance: Clear Liq.
   Odor: Essentially odorless
   Pour Point, C: -55
   Viscosity @ 25C, cps: 28
   Volatiles, %: 1.3
   Vapor Pressure, mm Hg @ 20C: 1 x 10 -9
   Flash Point (COC), C: 204

Morflex, Inc.: MORFLEX Plasticizers:

Morflex 150 (Dicyclohexyl Phthalate):
    CAS #: 84-61-7
    Molecular Weight: 330
    Molecular Formula: C20H26O4
    Morflex 150 is a solid at ambient conditions. It is used as
a plasticizer in heat-sealable films and as a co-plasticizer in
poly(vinyl chloride). Solidification from the liquid state is
time dependent, thus making it useful in thermosensitive adhesives.
    Specifications:
        Purity, wt%: 99.0 min
        Acidity, as phthalic acid, %: 0.014 max
        Color (1:1 in toluene), APHA: 100 max
        Moisture (K-F), %: 0.1 max
        Odor: Characteristic

Morflex 190 (Butyl Phthalyl Butyl Glycolate):
    CAS #: 85-70-1
    Molecular Weight: 336
    Molecular Formula: C18H24O6
    Morflex 190 has found utility as a plasticizer in food-
packaging material, PVC tubing and dental cushions. Other poss-
ible uses of Morflex 190 are found in the Code of Federal
Regulations Vol. 21.
    Specifications:
        Appearance: Clear liquid
        Color, APHA: 50 max
        Odor @ 25C: Mild, characteristic
        Moisture (K-F), %: 0.15 max
        Specific Gravity, 25/25C: 1.095-1.105
        Refractive Index, 25C/D: 1.487-1.491

Morflex 530 (Triisodecyl Trimellitate):
    CAS #: 36631-30-8
    Molecular Weight: 630
    Molecular Formula: C39H66O6
    Specifications:
        Assay, w/w, %: 99.0 min
        Acidity, as acetic acid, %: 0.02 max
        Moisture (K-F), %: 0.10 max
        Color, APHA: 100 max
        Specific Gravity @ 25/25C: 0.967-0.971
        Refractive Index @ 25C/D: 1.4820-1.4840

Morflex, Inc.: MORFLEX Plasticizers (Continued):

Morflex 560 (Tri-n-hexyl Trimellitate):
    CAS #: 1528-49-0
    Molecular Weight: 462
    Molecular Formula: C27H42O6
    The major use for tri-n-hexyl trimellitate is as a plastic-
izer for PVC and its copolymers. It provides the excellent
compatibility and efficiency properties of monomerics together
with the permanence properties of many polymerics.
    Vinyls plasticized with this ester find application in a
variety of markets including automotive and furniture upholstery, wire coatings and gasketing materials.
Specifications:
    Assay, w/w, %: 99.0 min
    Acidity, as acetic acid, %: 0.02 max
    Moisture (K-F), %: 0.10 max
    Color, APHA: 100 max
    Specific Gravity @ 25/25C: 1.010-1.020
    Refractive Index @ 25C/D: 1.4840-1.4860

Morflex 1129 (Dimethyl Isophthalate):
    CAS #: 1459-93-4
    Molecular Weight: 194
    Molecular Formula: C10H10O4
    One of the principal uses of Dimethyl Isophthalate is as
an intermediate in the synthesis of polyesters.
Specifications:
    Purity, wt%: 99.5 min
    Acidity, as acetic acid, %: 0.01 max
    Moisture (K-F), %: 0.10 max

Diethyl Isophthalate:
    CAS #: 636-53-3
    Molecular Weight: 222.2
    Molecular Formula: C12H14O4
Specifications:
    Assay (GC), %: 99.0 min
    Acidity, w/w, %: 0.005 max
    Moisture (K-F), %: 0.05 max

Dimethyl Adipate:
    CAS #: 627-93-0
    Molecular Weight: 174
    Molecular Formula: C8H14O4
Specifications:
    Clear colorless liquid
    Assay, GLC, %: 99.5 min
    Odor @ 25C: Mild, characteristic
    Color, APHA: 20 max
    Acidity, as adipic, %: 0.02 max
    Moisture (K-F), %: 0.10 max

BFGoodrich Kalama, Inc.: K-FLEX DE, DP, and 500:

K-Flex DE: K-Flex DP:
    Diethylene Glycol Dibenzoate   Dipropylene Glycol Dibenzoate
    Formula: (C6H5CO2CH2CH2)2O   Formula: (C6H5CO2CHCH3CH2)2O
        C18H18O5   C20H22O5
    Molecular Weight: 314   Molecular Weight: 342
    CAS Reg Number: 120-55-8   CAS Reg Number: 27138-31-4
        K-Flex 500 is a blend of K-Flex DE and K-Flex DP

Uses:
    K-Flex dibenzoate esters are widely used as plasticizers for
polyvinyl acetate emulsion systems. In these systems, K-Flex
dibenzoate esters reduce set times, increase open times, increase
film flexibility, reduce heat sealing temperatures, and improve
the dried film's resistance to oil, grease, and water. They also
serve to enhance solution viscosities, often eliminating the
need for other additives. K-Flex dibenzoate esters offer equal
or superior performance to other recognized adhesive plasticizers.
    Adhesives plasticized with K-Flex dibenzoate esters are used
in the following application areas: in the packaging industry
for carton sealing/forming; for book binding and labeling;
in textiles-for fibers as well as non-woven fabrics; and in
construction - to form decorative wall panels, window frames,
and other products as well as in the production of mastics and
caulking compounds. K-Flex plasticized adhesives are also used
in other applications such as furniture, luggage, shoes, and in
cigarette tipping.
    K-Flex dibenzoate esters are compatible with many other poly-
meric materials including ethylene vinyl acetate (EVA), polyvinyl chloride (PVC), styrene-butadiene rubber (SBR), ethyl cellulose, nitrocellulose, cellulose acetate butyrate, nitrile rubber,
and acrylics.
    K-Flex dibenzoate esters are primary plasticizers for poly-
vinyl chloride. They are highly solvating monomeric plasticizers
that decrease processing times and lower processing temperatures.
    K-Flex dibenzoate esters are often used to replace plasticiz-
ers such dibutyl phthalate and butyl benzyl phthalate where the
formulator wishes to replace phthalate containing materials.

K-Flex DE:
    5 mm Hg Boiling Point (C): 240
    Weight per Gallon @ 25C: 9.8#
    Viscosity (Centistokes) @ 30C: 43
K-Flex 500:
    5 mm Hg Boiling Point (C): 235
    Weight per Gallon @ 25C: 9.6#
    Viscosity (Centistokes) @ 30C: 52
K-Flex DP:
    5 mm Hg Boiling Point (C): 230
    Weight per Gallon @ 25C: 9.4#
    Viscosity (Centistokes) @ 30C: 67

Hatco: HATCOL Plasticizer Products List:

Hatcol plasticizers are manufactured to rigid quality control specifications using an exclusive refining process. Hatcol plasticizers are distinguished by their high purity, low odor and color, and resistance to degradation in heat aging tests.

The plasticizers listed below are available at competitive prices:

Phthalates:
    Hatcol 2923: Diisodecyl Phthalate
    Hatcol 2922: Ditridecyl Phthalate

Adipates:
    Hatcol 2910: Diisodecyl Adipate
    Hatcol 2906: Diisooctyl Adipate

Sebacates:
    Hatcol 3110: Di-2-Ethylhexyl Sebacate

Trimellitates:
    Hatcol TOTM: Tri-2-Ethylhexyl Trimellitate

Hatco Specialties:
    Hatcol 200: DOP replacement for PVC medical applications

Inolex Chemical Co.: LEXOLUBE Esters:

Lexolube 2X-114:
  Di 2-ethylhexyl Adipate
  Viscosity @ 25C (cSt): 12
  Viscosity @ 100C (cSt): 2.2
  VI: 120
  Smoke Point (C): 158

Lexolube 2X-130:
  Diisooctyl Adipate
  Viscosity @ 25C (cSt): 14
  Viscosity @ 100C (cSt): 2.7
  VI: 148
  Smoke Point (C): 160

Lexolube 2X-108:
  Diisodecyl Adipate
  Viscosity @ 25C (cSt): 24
  Viscosity @ 100C (cSt): 3.4
  VI: 148
  Smoke Point (C): 165

Lexolube 2X-109:
  Ditridecyl Adipate
  Viscosity @ 25C (cSt): 51
  Viscosity @ 100C (cSt): 5.5
  VI: 137
  Smoke Point (C): 180

Lexolube 2I-237:
  Tetraethylene Glycol Di-heptanoate
  Viscosity @ 25C (cSt): 15
  Viscosity @ 100C (cSt): 2.8
  VI: 149
  Smoke Point (C): 163

Lexolube 2N-237:
  Tetraethylene Glycol Di Caprylate/Caprate
  Viscosity @ 25C (cSt): 20
  Viscosity @ 100C (cSt): 3.4
  VI: 170
  Smoke Point (C): 156

Lexolube 2J-237:
  Tetraethylene Glycol Dicocoate
  Viscosity @ 25C (cSt): 36
  Viscosity @ 100C (cSt): 5.0
  VI: 188
  Smoke Point (C): 174

Rit-Chem Co., Inc.: RIT-CIZER #8 Liquid Plasticizer:
N-Ethyl o/p Toluenesulfonamide:

Rit-Cizer #8 is a liquid plasticizer of the sulfonamide group designed as a multi-purpose product. It is applied to various type of adhesives, paints, printing inks, epoxy resins, polyamide resins, phenolic resins, melamine resins and other resins.

Rit-Cizer #8 has many outstanding qualities which it imparts, such as:
1. Increased adhesion to adhesives and coatings
2. Increased flexibility at lower temperatures
3. Gloss improvement
4. Improves oil resistance
5. Improves stability at higher temperatures
6. Increases water resistance properties
7. Decreases abrasion loss
8. Prevents blotting & smudging of printing inks

Rit-Cizer #8 is used primarily to improve adhesion and promote flexibility of various resin systems. Since it is highly polar and readily plasticizes with most resins, its compatibility makes Rit-Cizer #8 extremely useful in formulations. It imparts resistance to greases, oils and solvents. Some of the resins in which Rit-Cizer #8 exhibits excellent compatibility and improves performance are:
Polyamide Resins
Nitrocellulose Lacquers
Polyvinyl Acetate & Ethyl Vinyl Acetate
Cellulose Acetate

FDA Data:
Rit-Cizer #8 can be used in certain indirect food contact applications.

Formula: $CH_3 C_6H_4 SO_2NH C_2H_5$
Molecular Weight: 199.2
Appearance: Light yellow, viscous liquid
Odor: Slight (characteristic)
Sales Specifications:
    Moisture (KF): 0.5% Max
    Color (APHA): 400 Max
    Ignition Residue: 0.03% Max
Physical Properties:
    Specific Gravity (50C): 1.180-1.195
    Freezing Point: OC Max
    Flash Point: 75C
    Boiling Point (760 mmHg): 340C

Solutia, Inc.: Plasticizers:

Santicizer 160:
   Plasticizer Excellence for High Solvent Action, Stain Resistance
   Butyl Benzyl Phthalate

   Santicizer 160 is an excellent plasticizer with strong solvent action on a wide variety of resins.
   Molecular Weight: 312
   Acidity (meq/100 gm. max): 0.37
   Appearance: Clear, oily liquid
   Color (APHA) (Max.): 40
   Moisture (KF in Methanol) %, max.: 0.15
   Odor: Slight, characteristic
   Refractive Index (@ 25C): 1.535-1.540
   Specific Gravity (25/25C): 1.115-1.123

Santicizer 261:
   Low Volatility, High Solvating Plasticizer
   Alkyl Benzyl Phthalate

   Santicizer 261 is a monomeric plasticizer that offers high performance, comparable to low molecular weight polymerics.
   Molecular Weight: 368
   Acidity (meq/100 gm. max): 0.37
   Appearance: Clear, oily liquid
   Color (APHA) (max.): 75
   Moisture (KF in Methanol) %, max.: 0.15
   Odor: Slight, characteristic
   Refractive Index (@ 25C): 1.523-1.529
   Specific Gravity (25/25C): 1.065-1.074

Santicizer 278:
   Monomeric Plasticizer Offering Permanance that Approximates Polymerics
   Alkyl Benzyl Phthalate

   Santicizer 278 is a high molecular weight benzyl phthalate that offers very low volatility and good permanence yet retains the aggressive solvating characteristics of the benzyl phthalates
   Molecular Weight: 455
   Acidity (meq/100 gm. max): 0.37
   Appearance: Clear, oily liquid
   Color (APHA) (max.): 175
   Moisture (KF in Methanol) % max.: 0.15
   Odor: Slight, characteristic
   Refractive Index (@25C): 1.516-1.520
   Specific Gravity (25/25C): 1.093-1.100

Solutia Inc.: Plasticizers (Continued):

DOA:
   Low-Temperature Plasticizer for PVC Resins and Synthetic
Rubbers
   Dioctyl Adipate/bis (2-Ethylhexyl) Adipate

   DOA is a high-quality plasticizer for imparting low-temp-
erature flexibility to polyvinyl chloride and synthetic rubber
compositions.
   Molecular Weight: 371
   Acidity (meq/100 gm. max): 0.25
   Appearance: Clear, oily liquid
   Color (APHA) (max.): 25
   Moisture (KF in Methanol) %, max.: 0.10
   Odor: Mild
   Refractive Index (@ 25C): 1.444-1.448
   Specific Gravity (25/25C): 0.921-0.927

Santicizer 97:
   Superior Low-temperature Flexibility with Reduced Volatility
for PVC Film, Sheet and Coatings
   Dialkyl Adipate

   Santicizer 97 is a plasticizer designed to give PVC film,
sheet and coatings extreme low-temperature flexibility.
   Molecular Weight: 370
   Acidity (meq/100 gm. max): 0.25
   Appearance: Clear, oily liquid
   Color (APHA) (max.): 50
   Moisture (KF in Methanol) %, max.: 0.10
   Refractive Index (@ 25C): 1.441-1.447
   Specific Gravity (25/25C): 0.916-0.924

Santicizer 141:
   FDA-regulated, Flame-retardant, Low-smoke, Low-temperature
Plasticizer
   2-Ethylhexyl Diphenyl Phosphate

   Santicizer 141 is an excellent, general-purpose plasticizer
for most commercial resins, including polyvinyl chloride and
its copolymers, cellulose nitrate, cellulose acetate-butyrate,
ethyl cellulose, polymethyl methacrylate and polystyrene.
   Molecular Weight: 362
   Phosphorus, %: 8.6 (Calc)
   Acidity (meq/100 gm. max): 0.20
   Appearance: Clear, oily liquid
   Color (APHA) (max.): 30
   Moisture (KF in Methanol) %, max.: 0.10
   Odor: Essentially odorless
   Refractive Index (@ 25C): 1.506-1.510
   Specific Gravity (25/25C): 1.085-1.091

Solutia Inc.: Plasticizers (Continued):

Santicizer 148:
    Low-smoke, Flame-retardant, Low-temperature Plasticizer with Low Volatility
    Isodecyl Diphenyl Phosphate

    Santicizer 148 is an efficient flame retardant, excellent for many commercial resins, particularly polyvinyl chloride and its copolymers, polyvinyl acetate and acrylics.
    Molecular Weight: 390
    Phosphorus, %: 7.9 (Calc)
    Acidity (meq/100 gm. max): 0.20
    Appearance: Clear, oily liquid
    Color (APHA) (Max.): 100
    Moisture (KF in Methanol) %, max.: 0.10
    Odor: Essentially odorless
    Refractive Index (@ 25C): 1.501-1.507
    Specific Gravity (25/25C): 1.061-1.071

Santicizer 2148:
    Very Low-smoke, Flame-retardant Plasticizer with Very Low Volatility and Outstanding Low-temperature Performance

    Santicizer 2148 is a low-smoke, flame-retardant alkyl aryl phosphate ester plasticizer with very low volatility, excellent low-temperature performance and good softening efficiency.
    Molecular Weight: Proprietary
    Phosphorus, %: Proprietary
    Acidity (meq/100 gm max): 0.20
    Appearance: Clear, oily liquid
    Color (APHA) (Max): 200
    Moisture (KF in Methanol) % max: 0.10
    Odor: Essentially odorless
    Refractive Index (@ 25C): 1.494-1.502
    Specific Gravity (25/25C): 1.028-1.044
    Density (@ 25C) ca lbs/gal: 8.65
    Crystallizing Point: 0C (32F)

Santicizer 143:
    Flame-retardant Plasticizer
    Proprietary Triaryl Phosphate Ester

    Santicizer 143 is a proprietary, modified triaryl phosphate ester, designed for use as a flame-retardant plasticizer for a variety of polymer systems.
    Phosphorus, %: 8.2 (Calc)
    Acidity (meq/100g) max.: 0.20
    Appearance: Clear, oily liquid
    Color (APHA) (max): 100
    Moisture % max.: 0.15%
    Refractive Index (@ 25C): 1.538-1.544
    Specific Gravity (25/25C): 1.142-1.156

Solutia, Inc.: Plasticizers (Continued):

Santicizer 154:
  Flame-retardant Additive and Plasticizer
  t-Butylphenyl Diphenyl Phosphate

  Santicizer 154 plasticizer is a triaryl phosphate ester with
flame-retardant properties suitable for use in a variety of
polymer systems.
  Molecular Weight (avg.): 371
  Phosphorus, %: 8.4 (Calc)
  Acidity (meq/100g.) max.: 0.20
  Appearance: Clear, mobile liquid
  Color (APHA) (max.): 100
  Moisture, % max.: 0.15
  Odor: Essentially odorless

Triphenyl Phosphate:
  Flame-retardant Plasticizer for Engineering Thermoplastics
and Cellulosics

  Triphenyl phosphate-a solid-is the most commonly used flame
retardant for cellulose acetate.
  Molecular Weight: 326
  Phosphorus, %: 9.5 (Calc.)
  Acidity (meq/100 gm. max): 0.10
  Appearance: White flakes
  Color (APHA) (max.): 20 (molten)
  Odor (max.): Very faint, aromatic

HB-40:
  Plasticizer Extender, Polymer Modifier, High-boiling Resin
Solvator
  Hydrogenated Terphenyl

  HB-40 plasticizer is a high-boiling aromatic which solvates
a variety of polymers, rubbers, asphalts and tars.
  Appearance: Clear, oily liquid
  Color (APHA) (max.): 450
  Moisture (KF in Methanol) ppm, max.: 150
  Odor: Faint, characteristic
  Refractive Index (@ 25C): 1.570-1.582
  Specific Gravity (25/15.5C): 1.001-1.009

Harwick Standard Distribution Corp.: Dover Chemical Corp.:
PAROIL And CHLOROFLO Liquid Chlorinated Paraffins:

Paroil 10:
    Color, Typical, Gardner 1933 Std.: 1
    Chlorine Content, % by weight: 40
    Specific Gravity @ 50/25C: 1.060
                 @ 25/25C: 1.080
    Pounds per Gallon: 8.9
    Viscosity: SUS @ 210F: 33
    Volatility, % Loss: 4 hrs @ 150C: 60.0

Chloroflo 40:
    Color, Typical, Gardner 1933 Std.: 2
    Chlorine Content, % by weight: 39
    Specific Gravity @ 50/25C: 1.085
                 @ 25/25C: 1.100
    Pounds per Gallon: 9.0
    Viscosity: SUS @ 210F: 63
    Volatility, % Loss: 4 hrs @ 150C: 5.0

Chloroflo 42:
    Color, Typical, Gardner 1933 Std.: 2
    Chlorine Content, % by weight: 40
    Specific Gravity @ 50/25C: 1.100
                 @ 25/25C: 1.120
    Pounds per Gallon: 9.3
    Viscosity: SUS @ 210F: 90
    Volatility, % Loss: 4 hrs @ 150C: 2.0

Paroil 140:
    Color, Typical, Gardner 1933 Std.: 1
    Chlorine Content, % by weight: 42
    Specific Gravity @ 50/25C: 1.150
                 @ 25/25C: 1.170
    Pounds per Gallon: 9.6
    Viscosity: SUS @ 210F: 150
    Volatility, % Loss: 4 hrs @ 150C: 1.5

Paroil 142A:
    Color, Typical, Gardner 1933 Std.: 1
    Chlorine Content, % by weight: 45
    Specific Gravity @ 50/25C: 1.195
                 @ 25/25C: 1.215
    Pounds per Gallon: 10.0
    Viscosity: SUS @ 210F: 200
    Volatility, % Loss: 4 hrs @ 150C: 1.5

Harwick Standard Distribution Corp.: Dover Chemical Corp.:
PAROIL and CHLOROFLO Liquid Chlorinated Paraffins:

Paroil 50:
    Color, Typical, Gardner 1933 Std.: 1
    Chlorine Content % by Weight: 50
    Specific Gravity @ 50/25C: 1.250
                     @ 25/25C: 1.270
    Pounds per Gallon: 10.3
    Viscosity: SUS @ 210F: 40
    Volatility, % Loss: 4 hrs @ 150C: 25.0

Paroil 152:
    Color, Typical, Gardner 1933 Std.: 1
    Chlorine Content % by Weight: 51
    Specific Gravity @ 50/25C: 1.250
                     @ 25/25C: 1.270
    Pounds per Gallon: 10.3
    Viscosity: SUS @ 210F: 70
    Volatility, % Loss: 4 hrs @ 150C: 3.5

Paroil 150A:
    Color, Typical, Gardner 1933 Std.: 1
    Chlorine Content % by Weight: 50
    Specific Gravity @ 50/25C: 1.250
                     @ 25/25C: 1.270
    Pounds per Gallon: 10.3
    Viscosity: SUS @ 210F: 450
    Volatility, % Loss: 4 hrs @ 150C: 1.5

Paroil 150LVA:
    Color, Typical, Gardner 1933 Std.: 2
    Chlorine Content % by weight: 48
    Specific Gravity @ 50/25C: 1.220
                     @ 25/25C: 1.240
    Pounds per gallon: 10.3
    Viscosity: SUS @ 210F: 250
    Volatility, % Loss: 4 hrs @ 150C: 1.4

Paroil 1061:
    Color, Typical, Gardner 1933 Std.: 1
    Chlorine Content % by Weight: 61
    Specific Gravity @ 50/25C: 1.370
                     @ 25/25C: 1.390
    Pounds per Gallon: 11.4
    Viscosity: SUS @ 210F: 72
    Volatility, % Loss: 4 hrs @ 150C: 11.0

Harwick Standard Distribution Corp.: Dover Chemical Corp.:
PAROIL and CHLOROFLO Liquid Chlorinated Paraffins (Continued):

Paroil 57/61:
   Color, Typical, Gardner 1933 Std.: 1
   Chlorine Content % by Weight: 59
   Specific Gravity @ 50/25C: 1.335
                    @ 25/25C: 1.355
   Pounds per Gallon: 11.1
   Viscosity SUS @ 210F: 63
   Volatility, % Loss: 4 hrs @ 150C: 14.0

Paroil 1650:
   Color, Typical, Gardner 1933 Std.: 1
   Chlorine Content % by Weight: 62
   Specific Gravity @ 50/25C: 1.390
                    @ 25/25C: 1.410
   Pounds per Gallon: 11.8
   Viscosity SUS @ 210F: 90
   Volatility, % Loss: 4 hrs @ 150C: 6.0

Paroil 170LV:
   Color, Typical, Gardner 1933 Std.: 2
   Chlorine Content % by Weight: 67
   Specific Gravity @ 50/25C: 1.500
                    @ 25/25C: 1.520
   Pounds per Gallon: 12.5
   Viscosity SUS @ 210F: 350
   Volatility, % Loss: 4 hrs @ 150C: 5.2

Paroil 170HV:
   Color, Typical, Gardner 1933 Std.: 2
   Chlorine Content % by Weight: 70
   Specific Gravity @ 50/25C: 1.520
                    @ 25/25C: 1.540
   Pounds per Gallon: 12.7
   Viscosity SUS @ 210F: 530
   Volatility, % Loss: 4 hrs @ 150C: 1.0

Unitex Chemical Corp.: UNIPLEX Plasticizers:

Uniplex FRP 44-57:
   A liquid flame retardant plasticizer containing bromine and chlorine compounds. Improves on low temperature brittleness and lower smoke than conventional brominated phthalate diesters used in wire and cable industry applications.
Applications: Flexible PVC, EPDM, and TPO applications in the
   wire and cable industries.
Features:
   Pale yellow liquid
   Improved flame retardant efficiency due to bromine/chlorine
synergy
   Superior electrical properties
   Lower smoke (NBS smoke chamber)
   Improved low temperature brittleness properties
Typical Properties:
   Total esters: 98% minimum
   Gardner Color: 3 max

Uniplex 80: CAS No. 77-93-0:
   An excellent plasticizer for food packaging due to its pharmacological safety, economy, performance and lack of odor.
   Has been accepted as a component of plastic food wraps for use with all foods by the Food and Drug Administration.
   Uniplex 80 is economical both on a cost and a performance basis, while exhibiting excellent compatibility with a wide range of polymers, cellulosics, and resins. Resins plasticized with Uniplex 80 exhibit improved light-fastness and low toxicity.
Physical Properties:
   Molecular Weight: 276.3
   Refractive Index, 25C: 1.440 (typical)
   Weight per Gallon, 25C: 9.48 lb.
   Boiling Point, 1 mm Hg: 127C
   Viscosity, 25C: 35.2 cps (typical)
   Pour Point: -50F
   Flashpoint, COC: 155C

Uniplex 82: CAS No. 77-89-4:
   An excellent plasticizer for food packaging due to its pharmacological safety, economy, performance and lack of odor.
   Has been accepted as a component of plastic food wraps for use with all foods by the Food and Drug Administration.
   Also accepted for use as a plasticizer in aerosol hair sprays and bandages by the Food and Drug Administration.
   Economical both on cost and a performance basis, while exhibiting excellent compatibility with a wide range of polymers and resins. Resins plasticized with Uniplex 82 exhibit excellent heat stability, and low toxicity.
Physical Properties:
   Molecular Weight: 318.3
   Refractive Index, 25C: 1.438 (typical)
   Weight per gallon, 25C: 9.47 lb.
   Boiling Point, 1 mm Hg: 132C

Unitex Chemical Corp.: UNIPLEX Plasticizers (Continued):

Uniplex 83: Tri-n-Butyl Citrate: CAS No. 77-94-7:
   A safe economical plasticizer approved by the U.S. Food and
Drug Administration (FDA) for use in both indirect and direct
contact food applications according to Code Federal Regulation
21:              175.105 (Adhesives)
   Uniplex 83 has been widely formulated/plasticize in poly-
vinyl chloride polymers and cellulosic resins as an "environmentally friendly" replacement to the phthalate plasticizers.
   The low odor and excellent color stability of Uniplex 83
make it an excellent general purpose plasticizer.
   Uniplex 83 is very effective as a defoamer in protein solutions. Uniplex 83 does not support fungal growth in resins.
   Other uses of Uniplex 83 include:
   * PVC flooring   * dairy product cartons
   * drink bottle caps   * food jar caps
   Uniplex 83 is economical both on a cost and a performance
basis, while exhibiting excellent compatibility with a wide
range of polymers and resins.

Uniplex 84: Acetyl Tributyl Citrate: CAS No. 77-90-7:
   Uniplex 84 is a safe economical plasticizer approved by
the U.S. Food and Drug Administration (FDA) for use in both
indirect and direct contact food applications according to
Code Federal Regulation 21.
   Uniplex 84 is used as the milling lubricant for aluminum
foil or sheet steel or for use in cans for beverage and food
products.
   Has been widely formulated in polyvinyl chloride polymers
for children's toys as a safe alternative to other widely used
plasticizers.
   The low odor and excellent color stability of Uniplex 84
make it an excellent general purpose plasticizer. Uniplex 84
is a good plasticizer for cellulose acetate films, replacing
dibutyl phthalate.
   Other uses of Uniplex 84 include:
   * aerosol hair sprays   * dairy product cartons
   * drink bottle caps   * food jar caps
   * solution coatings for foil and paper
   Uniplex 84 is economical both on a cost and a performance
basis, while exhibiting excellent compatibility with a wide
range of polymers and resins.
Physical Properties:
   Molecular Weight: 402.5
   Refractive Index 25C: 1.441 (typical)
   Weight per gallon, 25C: 8.74 lb
   Boiling Point, 1 mm Hg: 173C
   Viscosity, 25C: 25-35 cps
   Pour Point: -75C
   Flashpoint, COC: 204C

Unitex Chemical Corp.: UNIPLEX Plasticizers (Continued):

Uniplex 105: Diethyl Phthalate: CAS No. 84-66-2:
Specifications:
   Appearance: Clear Oily Liquid
   Color, (APHA): 10 Max
   Acidity (as Acetic acid): 0.007% Max
   Refractive Index, 25C: 1.499-1.501
   Assay, (% Diethyl Phthalate): 99% Min
   Moisture, Karl Fisher: 0.10% Max
Typical Properties:
   Specific Gravity, 20C: 1.118-1.122
   Odor: Essentially odorless
   Molecular Weight: 222

Uniplex 108: N-Ethyl o/p-Toluene Sulfonamide:
CAS No. 1077-5601 and 80-39-7:
   Uniplex 108 is one of the most widely compatible plasticizers.
It readily plasticizes nylon, other polyamides, shellac, cellulose acetate and protein materials. Uniplex 108 is outstanding
for making compounds resistant to oils, solvents and greases.
   For nylons, Uniplex 108 is one of the best plasticizers known.
However, unless FDA clearance is needed, Uniplex 214 is a more
economical choice. Uniplex 108 imparts toughness without
adversely affecting other properties. It lowers the melting point
and improves low temperature flexibility.
   In polyvinyl acetate adhesives, Uniplex 108's good grease
resistance and flexibility develops good tack for joining difficult surfaces, such as metal to rubber. In hot melts, Uniplex
108 improves flexibility for formulating book bindings and shoe
adhesives.
   In cellulose acetate formulations, Uniplex 108 imparts
excellent brilliance and gloss, while improving water and scuff
resistance, oil resistance, and gives heat and light resistance
to discoloration.
   In nitrocellulose lacquers, Uniplex 108 enhances adhesion,
flexibility, and resistance to water, oils, and greases.
   Uniplex 108 is approved by the U.S. Food and Drug Administra-
tion (FDA) according to Code of Federal Regulation 21.
Specifications:
   Appearance: Clear, oily liquid, with no suspended matter
   Acidity (meq/100g): 0.8 max
   Color (APHA): 200 max
   Odor: Characteristic
   Moisture: 0.5% max
   Assay (ortho/para isomers): 97% min
Properties:
   Molecular Weight: 199
   Refractive Index @ 25C: 1.535-1.545
   Specific Gravity @ 25C: 1.184-1.187
   Pour Point: -10C or +14F typical

Unitex Chemical Corp.: UNIPLEX Plasticizers (Continued):

Uniplex 110: Dimethyl Phthalate: CAS No. 131-11-3:
   Solvent and plasticizer for cellulose acetate-butyrate
compositions.
   Widely used as a solvent for organic catalysts.
   Excellent plasticizer and solvent for aerosol hair sprays.
Specifications:
   Appearance: Clear Liquid
   Color, (APHA): 10 Max
   Acidity (as phthalic acid): 0.1% Max
   Refractive Index, 25C: 1.512-1.514
   Assay, (% Dimethyl Phthalate): 99% Min
   Moisture: 0.1% Max

Uniplex 125-A: Dioctyl Adipate [Di(2-ethyl hexyl) Adipate]:
CAS No. 103-23-1:
   A versatile plasticizer that is compatible with many polymers,
including polyvinyl chloride, polyvinyl acetate, polyvinyl butyral, ethyl cellulose, nitrocellulose, and cellulose acetate
butyrate.
   Uniplex 125-A is approved as an indirect food addition under
21CFR 175.105, 177.1210 and 177.2600.
   Provides superior flexibility to vinyl products at low
temperatures compared to dioctyl phthalate.
   In comparison to dioctyl phthalate, Uniplex 125-A is advant-
ageous in preparing vinyl dispersions with low initial and
stable viscosity.
   The electrical resistivity and low dissipation factor of
Uniplex 125-A make it especially useful in electrical insulation
applications. Other suggested applications are polyvinyl chloride
film, polyvinylidene chloride film, and adhesive and coatings
formulations.
Specifications:
   Appearance: Clear Liquid
   Acid No. (mg KOH/g): 0.15 max(equivalent to 0.02% Adipic Acid)
   Color (APHA): 25 max
   Refractive Index (25C): 1.444-1.448
   Specific Gravity (25C): 0.922-0.927
   Assay: 99.0 min
   Moisture, %: 0.1 max

Unitex Chemical Corp.: UNIPLEX Plasticizers (Continued):

Uniplex 150: Dibutyl Phthalate: CAS No. 84-74-2:
Uniplex 150 is an excellent general purpose plasticizer that is compatible with cellulose acetate, cellulose acetate butyrate, cellulose nitrate, ethyl cellulose, polymethyl methacrylate, polystyrene, polyvinyl butyral, vinyl chloride, and vinyl chloride acetate. It also is an excellent plasticizer for thermosetting resins such as urea-formaldehyde, melamine-formaldehyde, phenolics, and others.
Uniplex 150 is approved under Code of Federal Regulations 21.
Uniplex 150 is suggested as a solvent/plasticizer in fingernail polish, nail polish remover, hair sprays, organic peroxide catalysts, adhesive, and various coatings.
Specifications:
    Appearance: Clear liquid
    Acidity as Phthalic Acid: 0.01% max
    Color, APHA: 25 max
    Refractive Index, 25C: 1.4905 typical
    Assay: 99.0% min
    Moisture, K.F.: 0.1% max

Uniplex 155: Di-Isobutyl Phthalate: CAS No. 84-69-5:
Uniplex 155 is a general purpose plastcizer that is widely compatible with both thermoplastic and thermosetting resins.
Uniplex 155 is approved under Code of Federal Regulations 21.
Uniplex 155 is suggested as a solvent/plasticizer in cellophane, resin coated sand for foundry casting and organic peroxides.
Specifications:
    Appearance: Clear Liquid
    Acidity (as phthalic acid): 0.01% max
    Color (APHA): 25 max
    Assay: 99.0% min
    Moisture: 0.1% max

Uniplex 165: Diisobutyl Adipate: CAS No. 141-04-8:
Uniplex 165 is a safe, effective plasticizer for imparting low temperature flexibility to various polymers, with a good balance of solvation and low temperature properties. It has excellent compatibility with cellulose acetate butyrate, cellulose nitrate, vinyl acetate, vinyl chloride, and vinyl chloride acetate.
Uniplex 165 is approved under Code of Federal Regulations 21.
Suggested as a solvent/plasticizer in fingernail enamel, nail enamel remover, hair sprays, adhesives, and coatings. It is particularly recommended for plasticizing resinous and polymeric coatings that are intended for direct contact with foodstuffs.
Specifications:
    Appearance: Clear liquid, no suspended material
    Assay, as Diisobutyl Adipate: 99.0% min
    Color (APHA): 25 max
    Moisture, Karl Fischer: 0.10% max
    Specific Gravity, 25/25C: 0.946-0.953

Unitex Chemical Corp.: UNIPLEX Plasticizers (Continued):

Uniplex 171: o,p-Toluene Sulfonamide: CAS No. 70-55-3/88-19-7
An excellent plasticizer for thermosetting and thermoplastic resins. It is regulated by FDA under Section 175.105 of the Federal Register.
Imparts good gloss and wetting action to melamine, urea and phenolic resins, improving mixing of filled product.
Increases hardness of moldings made with urea resins.
Increases stability and compatibility of melamine resins; better strength, abrasion resistance and higher gloss are obtained.
Reduces viscosity and improves flow of phenolic resins.
Increases flexibility with only minor decreases in tensile strength in nylon and other polyamide resins.
Properties:
Molecular Weight: 171
Appearance: Fine white to light cream solid particles
Moisture (KF in Methanol): 1.0% max (Typical analysis 0.05%)
Odor: Essentially odorless
Flash Point (COC): 420F
Ortho Content: 30+-5%
Para Content: 70+-5%

Uniplex 214: N-butylbenzene Sulfonamide: CAS No. 3622-84-2:
Uniplex 214 (N-Butylbenzene sulfonamide or BBSA) is a liquid sulfonamide plasticizer for a number of resins, such as poly-acetals, polycarbonates, polysulfones, and polyamides, especially Nylon-11 and Nylon-12. The addition of this plasticizer contributes the following properties:
* Easier removal from molds
* Easier machining
* Better finish
* Good heat stability to 80C
* Better shape stability due to reduced water absorption
Uniplex 214 is the preferred primary plasticizer for poly-amide resins, greatly enhancing low temperature flexiblity.
Uniplex 214 may be added either pre- or post-polymerization for Nylon-11 or 12, and in the extrusion of other polyamides (e.g. Nylon 6, Nylon 66). It is typically used at levels of 5-20%, based on the weight of the polymer.
Uniplex 214 is used in such applications as molded polyamide parts, nylon fishing line, nylon-based adhesives for nonwoven interlinings, auto fuel line, coil air hoses, and other high performance applications.
Molecular Weight: 213.3 g/mole
Empirical Formula: C10H15NO2S
Specifications:
Appearance: Clear liquid
Odor: Mild
Color, Pt-Co Units: 50 max
Water Content %: 0.1 max
Basicity, meq/kg: 4 max
Assay: 99% min

Unitex Chemical Corp.: UNIPLEX Plasticizers (Continued):

Uniplex 225: N-(2-hydroxypropyl)benzenesulfonamide:
CAS No. 35325-02-1:
    Uniplex 225 is a highly effective plasticizer for polyamide
and polyurethane polymers with excellent anti-static properties.
    Also suggested for use in polyacrylic and cellulose ester.
    Exhibits excellent resistance to extraction by water and dry
cleaning solvents.
Properties:
    Appearance: Clear Viscous Liquid
    Moisture, Karl Fisher: 5.0-7.0%
    Color, APHA: 50 max
    Refractive Index, 25C: 1.529-1.535
    Solids Content: 93-96%

Uniplex 250: Dicyclohexyl Phthalate: CAS No. 84-61-7:
    Uniplex 250 is FDA approved under 21CFR 175.105, 21CFR176.170
& 21CFR 177.1200.
    Uniplex 250 is a heat activated plasticizer used in heat
seal applications such as food wrappers, food labels, pharma-
ceutical labels and other applications where a delayed heat
activated adhesive is required.
    Uniplex 250 is used in printing ink formulations to improve
adhesion and water resistance when applied to paper, vinyl,
textile and other substrates.
Specifications:
    Appearance: White crystalline solid lumps or granules
    Acidity (as phthalic acid): 0.01% max
    Color (APHA-1:1 isopropanol): 50 max
    Melting Point (C): 63-65
    Assay: 99% min

Uniplex 260: Glyceryl Tribenzoate: CAS No. 614-33-5:
    Uniplex 260 is a solid polyol benzoate modifier for a wide
range of polymers. Considered as a very safe plasticizer.
    Uniplex 260 is approved by the FDA according to Code of
Federal Regulations 21.
    Uniplex 260 is especially recommended for use in heat seal
applications. It is blended into polyvinyl acetate based-adhe-
sives at a temperature below the melting point.
    May be used in cellophane coatings to improve heat-seal
properties, without affecting color and clarity of the film.
    An outstanding plasticizer for nitrocellulose coatings where
it promotes heat-sealability and adhesion, creates a moisture
barrier and resists oil/water extraction. These properties are
of great value in heat seal coatings, lacquers and films.
    An excellent non-formaldehyde resin substitute and plastic-
izer in nail lacquer formulations. It imparts superior durability
and aesthetic properties.
    Other suggested applications for Uniplex 260 are:
    * metallic and pigmented surface coatings
    * extrusion and injection molding processing aids   * polishes
    * cleaning material for hot melt roll coaters

Unitex Chemical Corp.: UNIPLEX Plasticizers (Continued):

Uniplex 400: Polypropylene Glycol Dibenzoate: CAS No. 72245-46-6:
Uniplex 400 is a low volatility, moderately solvating plasticizer. An efficient, cost/performance alternative to higher
alkyl benzyl phthalate plasticizers.
Particularly useful in premium sealant applications. The low
volatility characteristics are of particular interest to formula-
tors of polysulfide and polyurethane sealants.
Also useful in aqueous emulsion systems such as caulks and
polyvinyl acetate adhesives. Uniplex 400 may be used to formulate
adhesives that need to provide good adhesion between porous and
polyolefin surfaces. In caulk formulations, the same excellent
performance of other benzoate plasticizers can be expected in
addition to providing lower volatility.
An excellent plasticizer for acrylic coatings and acrylic
foam coatings. Uniplex 400 is acrylic foam is more effective than
higher alkyl benzyl phthalates and tends to volatilize less than
those low volatility type phthalates.
Uniplex 400 is compatible with the following polymers or
polymer types (the list is not all inclusive):

Polyvinyl Acetate	Ethylene/Vinyl Acetates
Polyurethane	Polysulfides
Polyvinyl chloride	Ethylene/Polyvinyl chloride
Nitrocellulose	Acrylates
Polystyrene	Vinyl acrylates

Specifications:
  Appearance: Clear, light yellow liquid, free of suspended
             matter
  Odor: Characteristic
  Total Esters, G.C.: 98% min (as benzoate esters)

Uniplex 413: Substituted Benzene Sulfonamide:
(EPA Accession No. 51253):
Uniplex 413 is a high molecular weight sulfonamide that is
used as a plasticizer and nucleating agent in PET resins.
Used at levels of 4-10 percent on the weight of the PET
resin, Uniplex 413 greatly reduces molding (processing) times.
Uniplex 413 offers the advantages of:
1) lowering processing cost
2) enhances impact resistance
3) low vaporization
4) improves mold releases
5) easily added to the PET resin in single or double screw
   extruders
Specifications:
  Appearance: Yellowish soft particles (Clear liquid when
             molten)
  Color, APHA: 400 max (1:1 in Xylene)
  Acidity, meq/100 g: 0.25 max

Unitex Chemical Corp.: UNIPLEX Plasticizers (Continued):

Uniplex 512: Neopentyl Glycol Dibenzoate: CAS No. 4196-89-8:
Uniplex 512 is a low melting (49C) waxy solid plasticizer.
The product is used predominantly as thermoplastic processing
aid. These uses include:
* Incorporation in polycarbonate or talc-filled polypropylene
  to improve injection molding by cycle-time reduction.
* Incorporation in rigid vinyl compounds as an extrusion aid.
* Incorporation in thermoplastic polyesters (e.g. PBT, PCT,
  PET) to aid crystal growth during nucleation.
Uniplex 512 is an effective plasticizer for coatings appli-
cation, particularly in acrylic-based coatings where excellent
ultraviolet light stability is required.
Specifications:
Assay (Benzoate Esters), %: 98.0 min
Acidity (as Benzoic Acid), %: 0.1 max
Color, Pt-Co Units (20% in 1/1 Acetone/Xylene): 100 max
Hydroxyl Number, Mg KOH/g: 8.0 max
Water Content, %: 0.1 max

Uniplex 540: Applications in Polyvinyl Chloride:
Uniplex 540 is a high molecular weight monomeric tetraester
of pentaerythritol. It functions as a high performance plasticizer for polyvinyl chloride with outstanding heat resistance
and retention of elongation under conditions of elevated temperature. Uniplex 540 has additional benefits in vinyl plastics
because it combines ease of processing due to its low viscosity
with superior performance and oil resistance properties usually
associated only in various polymeric plasticizers.
Product Specifications:
Acid Number, mgKOH/g sample: 0.5 max
APHA Color: 150 max
Hydroxyl Number, mgKOH/g sample: 6 max
Water Content, %: 0.1 max
PVC Formulations:
PVC formulations plasticized with Uniplex 540 exhibit outstanding permanence, excellent heat and light stability, and
good insulating electrical properties. This combination of
properties is useful in rigorous thin-wall electrical wiring
applications, such as 105C wiring, and Government specification
cable construction, such as MIL-W-168780 and MIL-W-5086A, and
plenum wire and cable. Such wire coatings retain greater than
90% of their original elongation after four days of aging at
136C.

Unitex Chemical Corp.: UNIPLEX Plasticizers (Continued):

Uniplex 546: Tri(2-ethyl hexyl) Trimellitate: TOTM:
Uniplex 546 is an excellent low volatility plasticizer for polyvinyl chloride. It is particularly recommended for use in PVC wire insulation.
Specifications:
Clear liquid, free of foreign matter
Odor: Mild, characteristic
Assay: 99.0% min
Acidity, as Trimellitic Acid, wt%: 0.02 max
Color, APHA: 100 max, typical 70 or less
Water Content, wt%: 0.1 max
Specific Gravity, 20/20C: 0.989-0.995

Uniplex 552: Pentaerythritol Tetrabenzoate: CAS No. 4196-86-5:
Uniplex 552 is a solid plasticizer that can be formulated into adhesives intended for heat seal applications. It is approved by FDA under 21CFR 175.105.
Uniplex 552 is recommended for use in applications requiring higher heat activation temperatures that can be obtained with Uniplex 250 or Uniplex 260.
Specifications:
Appearance: Off-white Crystalline Particles (lumps or granules)
Acid No. (mg KOH/gm): 0.28 max
Color (APHA, 1:1 in Xylene): 100 max
Refractive Index (Molten 50C): 1.569-1.572

Uniplex 809: PEG Di-2-Ethylhexoate: CAS No. 9004-93-7:
Uniplex 809 is a high molecular weight liquid plasticizer, having wide compatibility with a variety of polymers. Its low volatility and excellent heat resistance make it especially useful in engineering plastics, such as polyester and polyamide resins.
Used at levels of 4-10% on the weight of the PET resin, Uniplex 809 greatly reduces molding (processing) times.
Uniplex 809 offers the advantages of:
1. Lowering processing cost
2. Enhancing impact resistance
3. Low vaporization
4. Improving mold releases
5. Easily added to the PET resin in single or double screw extruders.
Specifications:
Appearance: Clear to slightly hazy, oily liquid
Acid value (mg KOH/g): 1.0 max
Hydroxyl Value (mg KOH/g): 2.5 max
Sum of Acid Value and Hydroxyl Value: 3.0 max
Color, APHA: 200 max
Moisture, %: 0.10 max

Unitex Chemical Corp.: UNIPLEX Plasticizers (Continued):

Uniplex 810: PEG-Di Laurate: CAS No. 9005-02-1:
Uniplex 810 is a high molecular weight liquid plasticizer, and nucleating agent, having wide compatibility with a variety of polymers. Its low volatitility and excellent heat resistance make it especially useful in engineering plastics, such as poly-ester and polyamide resins.

Used at levels of 4-10% on the weight of the PET resin, Uniplex 810 greatly reduces molding (processing) times.

Uniplex 810 offers the advantages of:
1. Lowering processing cost
2. Enhancing impact resistance
3. Low vaporization
4. Improving mold releases
5. Easily added to the PET resin in single or double screw extruders.

Specifications:
Appearance: Grease below melting point, clear oily liquid above melting point
Acid Value (mg KOH/g): 5.0 max
Hydroxyl Value (mg KOH/g): 8.0 max
Color Gardner: 4 max
Moisture %: 0.10 max

Witco Corp.: Additives for Vinyl: Plasticizers:

Because plasticizers must be carefully chosen to achieve the most desirable balance of performance characteristics in particular compounds, a most valuable advantage of the Drapex line is its broad range of products.

Drapex epoxy plasticizers, used as co-stabilizers with Mark mixed-metal stabilizers, provide excellent heat and light stability.

DRAPEX Epoxy Plasticizers:

Witco Drapex epoxy plasticizers are the result of years of research, testing, and experience in the field. Drapex 6.8 and 10.4 epoxidized oils, and Drapex 4.4 epoxidized ester make important improvements in the appearance and durability of vinyl end products.

Their exceptional contribution to heat and light stability comes from their ability to act synergistically with primary stabilizers, particularly barium-zinc and calcium-zinc mixed-metal types, to achieve results that cannot be obtained with stabilizers alone.

Stability is increased during processing and in the final product where the useful life is greatly extended. Outdoor weathering tests Witco conducted in Arizona show that a Drapex epoxy plasticizer can triple or quadruple the life of a flexible vinyl compound.

You can choose and balance for a wide variety of characteristics, including excellent solvating action, low volatility, low-temperature flexibility and permanence. Two Drapex epoxy plasticizers, Drapex 6.8 and Drapex 10.4, are sanctioned by the FDA for certain food contact applications.

Section XVI
Processing Aids

BYK-Chemie USA: Anti-Separation Additives:

BYK Wetting and Dispersing Additives for Homogenization, Stabilization and Fiber wetting in Polyester Molding Compounds and Pultrusion.

Optimal Paste Manufacture:
* *Better reproducibility from batch to batch
* *Improved paste homogeneity
* *Anti separation of LS or LP components

Advantages during Molding:
* *Controlled shrinkage
* *Less stickiness to the film during removal

Advantage during Compound Manufacture:
* *Improved fiber wetting in HMC and pultrusion

Final Part Improvements:
* *Improved homogeneity
* *More homogeneous color with pigmented parts

BYK-W 972:
An additive to prevent phase separation and improve color homogeneity in LS systems.

BYK-W 973:
Cost effective alternative to BYK-W 972.

BYK-9075:
Solvent free additive to prevent separation and improve color homogeneity in LS or LP systems where residual solvent must be avoided.

Typical Data:

BYK-W 972:
Density at 20C/68F, lbs/gal: 8.51
Flash Point C/F: 38/100
Refractive Index: 1.438
Amine Value mg KOH/g: 11
Nonvolatile matter %: 30
Appearance: Light yellow liquid

BYK-W 973:
Density at 20C/68F, lbs/gal: 8.26
Flash Point C/F: 23/73
Refractive Index: 1.474
Amine Value mg KOH/g: 18
Nonvolatile matter %: 34
Appearance: Light yellow liquid

BYK-9075:
Density at 20C/68F, lbs/gal: 9.43
Flash Point C/F: >100/>212
Refractive Index: 1.535
Amine Value mg KOH/g: 12
Nonvolatile matter %: 97
Appearance: Light yellow liquid

BYK-Chemie USA: BYK-S 740, BYK-S 750: BYK Styrene Emission Suppressants for Unsaturated Polyester-, DCPD- and Vinyl Ester Resins:

BYK Emission Suppressants give:
    *Reduction of styrene monomer emissions 70-90%
    *No reduction of interlaminar adhesion
    *Easy procesing, no melting required
    *No negative influence on gel time or cure

BYK-S 740 was designed as a highly effective styrene suppressant in orthophthalic polyester resins in open mold applications.
BYK-S 740 also works in isophthalic polyester resins, depending on the resin polarity, the emission suppression is not so large in comparison to the effect on orthophthalic resins.

Which Additive to use for which Resin?
BYK-S 740:
    Orthophthalic resins: 0.5-1.0%
    Isophthalic resins: 0.5-1.0%

BYK-S 750:
    DCPD containing resins: 0.3-1.0%
    Isophthalic resins: 0.3-1.0%
    Vinyl ester resins: 0.3-1.0%

BYK-S 750 was developed specifically as a styrene suppressant for DCPD resins and DCPD blended resins in which BYK-S 740 is too soluble to form a surface film. It suppresses the emissions without adverse side effects on interlaminar adhesion.
It is also very effective in vinyl ester and isophthalic resins.

BYK-S 740:
    Solution of hydroxypolyesters with paraffin wax
    Weight per U.S gallon: lb/gal: 7.09
    Flashpoint: C/F: 57/135
    Refractive index: 1.457
    Acid value: mg KOH/g: <7
    Viscosity at 60C: 16.5
    Appearance: slightly brownish paste

BYK-S 750:
    Combination of wax with polar components
    Weight per U.S. gallon: lb/gal: 7.01
    Flashpoint: C/F: >100/>212
    Appearance: white paste

Croda Oleochemicals: CRODAMIDE Polymer Processing Additives:

Crodamides find a wide range of applications and are particularly used for solving issues with polymer processing and end use.

Primary Crodamide specifications:
Crodamide:
E:
    Erucamide
    Acid Value (max): 10
    Melting point, C: 75-80
    Colour Gardner (max): 10
ER:
    Refined Erucamide
    Acid Value (max): 1
    Melting point, C: 78-81
    Colour Gardner (max): 2
O:
    Oleamide
    Acid Value (max): 10
    Melting point, C: 66-72
    Colour Gardner (max): 10
OR:
    Refined oleamide
    Acid Value (max): 1
    Melting point, C: 70-73
    Colour Gardner (max): 2
ORX:
    Ref. heat stable oleamide
    Acid Value (max): 1
    Melting point, C: 71-76
    Colour Gardner (max): 2
ORV:
    Ref. vegetable oleamide
    Acid Value (max): 1
    Melting point, C: 70-75
    Colour Gardner (max): 2
S:
    Stearamide
    Acid Value (max): 10
    Melting point, C: 96-102
    Colour Gardner (max): 10
SR:
    Refined stearamide
    Acid Value (max): 5
    Melting point, C: 96-102
    Colour Gardner (max): 5
BR:
    Refined behenamide
    Acid Value (max): 3
    Melting Point, C: 102-112
    Colour Gardner (max): 4

**Croda Oleochemicals: CRODAMIDE Polymer Processing Additives
  (Continued):**

Secondary and bis Crodamide specifications:
Crodamide:
203:
  Oleyl palmitamide
  Acid Value (max): 2
  Melting point C: 60-66
  Amine value (max): 2
  Colour Gardner (Max): 3

212:
  Stearyl erucamide
  Acid Value (max): 2
  Melting point C: 70-75
  Amine value (max): 2
  Colour Gardner (max): 5

EBS:
  Ethylene bis-stearamide
  Acid Value (max): 5
  Melting Point C: 140-145
  Amine value (max): 5
  Colour Gardner (max): 5

EBO:
  Ethylene bis-oleamide
  Acid Value (max): 5
  Melting point C: 115-120
  Amine value (max): 5
  Colour Gardner (max): 10

DD8001:
  Proprietary blend
  Acid Value (max): 10
  Melting Point C: 80-90
  Amine value (max): 0-20
  Colour Gardner (max): 10

Dow Chemical Co.: Dow Industrial E-Series Polyglycols:

Industrial-grade E-Series Polyethylene Glycols (PEGs):
Dow E-series polyethylene glycols are polymers of ethylene oxide. As molecular weight increases, viscosity and freezing points increase, while solubility in water decreases. The molecular weight of the E-series polyglycols ranges from 200 to 8000. They are available in liquids (E200 to E600), waxy solids (E900 to E1450), and prills (E3350 to E8000). Dow Polyglycols E900 through E8000 are also available in heated molten form (bulk only). The E-series products can be blended together by users to produce materials with a particular viscosity or texture.

Rubber and Plastics:
In the rubber industry, polyglycols are used as components of clay-containing rubber formulations to control cure rates. Polyglycols are added to thermoplastics to improve flexibility as well as to impart antistatic properties to improve the processing of plastic pellets or resins. Hot baths of polyglycols also act as heat transfer fluids in plastic thermoforming. Polyglycols are used as lubricants to facilitate removal of vulcanized products from molds.

In other rubber and plastics applications, polyglycols offer lubricity, hygroscopicity, water solubility, and film-forming capabilities, and they are non-varnishing at high temperatures.

Typical Physical Properties:
Polyethylene Glycol E-Series:
CAS #25322-68-3:
E200:
    Average Molecular Weight: 200
    Average Freezing Point, C: Super Cools
    Average Viscosity, Centistokes: 32F: 187
    Average Viscosity, Centistokes: 77F: 40
    Average Viscosity, Centistokes: 100F: 23
    Flash Point PMCC, F: 340
    Refractive Index at 25C: 1.459
    Specific Gravity 25/25C: 1.124
    Density Lbs/Gal at 25C: 9.35
    Viscosity Index: 111
    Specific Heat Cal/g/C at 25C: 0.524
    CTFA Nomenclature: PEG-4
E300:
    Average Molecular Weight: 300
    Average Freezing Point, C: -10
    Average Viscosity, Centistokes: 32F: 343
    Average Viscosity, Centistokes: 77F: 69
    Average Viscosity, Centistokes: 100F: 36
    Flash Point PMCC, F: >400
    Refractive Index at 25C: 1.463
    Specific Gravity 25/25C: 1.125
    Density Lbs/Gal at 25C: 9.36
    Viscosity Index: 118
    Specific Heat Cal/g/C at 25C: 0.508
    CTFA Nomenclature: PEG-6

**Dow Chemical Co.: Dow Industrial E-Series Polyglycols
(Continued):**

Polyethylene Glycols E-Series (Continued):
E400:
    Average Molecular Weight: 400
    Average Freezing Point, C: +6
    Average Viscosity, Centistokes: 77F: 90
    Average Viscosity, Centistokes: 100F: 49
    Average Viscosity, Centistokes: 210F: 7.4
    Flash Point PMCC, F: >450
    Refractive Index at 25C: 1.465
    Specific Gravity 25/25C: 1.125
E600:
    Average Molecular Weight: 600
    Average Freezing Point, C: +22
    Average Viscosity, Centistokes: 77F: 131
    Average Viscosity, Centistokes: 100F: 72
    Flash Point PMCC, F: >450
    Refractive Index at 25C: 1.466
    Specific Gravity 25/25C: 1.126
E900:
    Average Molecular Weight: 900
    Average Freezing Point, C: 34
    Average Viscosity, Centistokes: 100F: 100
    Flash Point PMCC, F: >450
    Specific Gravity 25/25C: 1.204
E1000:
    Average Molecular Weight: 1000
    Average Freezing Point, C: 37
    Average Viscosity, Centistokes: 210F: 18
    Flash Point PMCC, F: >450
    Specific Gravity 25/25C: 1.214
E1450:
    Average Molecular Weight: 1450
    Average Freezing Point, C: 44
    Average Viscosity, Centistokes: 210F: 29
    Flash Point PMCC, F: >450
    Specific Gravity 25/25C: 1.214
E3350:
    Average Molecular Weight: 3350
    Average Freezing Point, C: 54
    Average Viscosity, Centistokes: 210F: 93
    Flash Point PMCC, F: >450
    Specific Gravity 25/25C: 1.224
E4500:
    Average Molecular Weight: 4500
    Average Freezing Point, C: 58
    Average Viscosity, Centistokes: 210F: 180
    Flash Point PMCC, F: >450
E8000:
    Average Molecular Weight: 8000
    Average Freezing Point, C: 60
    Average Viscosity, Centistokes: 210F: 800
    Flash Point PMCC, F: >500

Dow Corning: DOW CORNING MB40-006 Masterbatch:

Type: Ultra-high molecular weight siloxane polymer dispersed in
    polyoxymethylene (acetal) copolymer
Physical Form: Solid pellets
Special Features: Imparts benefits such as processing improve-
    ments and modified surface characteristics
Primary Use: Additive in acetal-compatible systems

Dow Corning MB40-006 Masterbatch is a pelletized formulation
containing 40 percent of an ultra-high molecular weight siloxane
polymer dispersed in polyoxymethylene (acetal) copolymer. It
is designed to be used as an additive in most acetal-compatible
systems to impart benefits such as processing improvements and
modification of surface characteristics.
Siloxane plastic additives have been used for several years
to improve the lubricity and flow of thermoplastics. They are
effective in this role although some difficiulties have been
experienced in the incorporation of liquids into thermoplastic
melts without use of specialized equipment. It has also been
difficult to produce masterbatches with greater than 20 percent
siloxane because of processing difficulty and bleed problems. The
Dow Corning MB series of siloxane masterbatches addresses this
problem by supplying a high concentration of an ultra-high
molecular weight siloxane as a dispersion in a dry pellet
form in a range of thermoplastics.

Benefits:
When added to acetal or similar thermoplastics at 0.1 to 1.0
percent, improved processing and flow of the resin is expected,
including better mold filling, less extruder torque, internal
lubrication, mold release, faster throughput, and less warpage
of the molded part. At higher siloxane addition levels, 1 to 5
percent siloxane, improved surface properties are expected,
including lubricity, slip, lower coefficient of friction, and
greater mar and abrasion resistance. The Dow Corning MB series
of solid additives is expected to give improved benefits compared to conventional lower molecular weight siloxane additives,
e.g., less screw slippage, improved release, a lower coefficient
of friction, and a broader range of performance capability.

Typical Properties:
    Appearance: Off-white pellets
    Siloxane Content, percent: 40
    Organic Resin: Polyoxymethylene (acetal) copolymer, MI 27
    Suggested Use Levels, Percent: 0.2 to 10 (0.1 to 5% siloxane)

Dyneon LLC: DYNAMAR Polymer Processing Additive:

Dynamar PPAs (Polymer Processing Additives) are designed to improve the processing of polyolefins. They are generally used at levels of 400 to 1000 ppm. In order for the process aids to function properly at these low levels, they must be well dispersed. The use of PPA concentrates will enable good dispersion and improved accuracy of addition amount.

For Use in:

Polyethylenes typically with a melt index less than or equal to 1
    *LLDPE (Linear Low Density Polyethylene)
    *HDPE (High Density Polyethylene)

Applications:

Extrusion processes where the shear rate is typically less than or equal to 2000 sec-1
    *Blown film   *Monofilament
    *Cast film   *Pipe/tubing
    *Extrusion Blow Molding   *Wire & Cable

Typical Use Levels:

Masterbatches are generally produced and used in concentrations of less than 10%, with 2-3% being preferred. These masterbatches are mixed in the polyethylene to produce levels of 400 to 1000 ppm (0.04 to 0.1 percent by weight) Dynamar PPA.

Benefits:

Increase Extrusion Output:

Increases in extrusion output of over 40% at equivalent die pressure have been observed with some resin compounds on a laboratory blown film line.

When polyethylene containing PPA is introduced into the extruder, the PPA will coat or condition the die wall. As the die wall is conditioned, there will be a reduction in back pressure, torque, and apparent viscosity. Increasing the extruder screw rpm to bring the back pressure up to the original reading will result in an increase in output.

The specific output increase will vary depending upon several factors including the type of process equipment used, the additive formulation, and the resin melt index and rheology.

Improve Dart Impact Resistance:

The use of PPA may eliminate melt fracture and surface defects which may result in an improvement in the dart impact resistance.

Process LLDPE with Dynamar PPA on LDPE Equipment:

Processors have successfully produced LLDPE blown film using equipment designed for processing LDPE.

Additional Benefits:

*PPA will allow processing of lower melt index, tougher, stronger resins. Low MI resins can be used to produce films and other articles at lower gauge with equivalent strength, saving expense

*Allows the processing to increase the amount of LLDPE (up to 100%) in blends with LDPE, in order to improve mechanical properties.

*Reduce or eliminate die drool or die build up

*May reduce power consumption

*May allow processing at lower temperatures

*Complies with FDA requirements for polyolefin food contact applications (usage rates up to 2000 ppm)

Eeonyx Corp.: EEONOMER Conductive Additives:

Eeonomer is:
  *Electrically and thermally conductive additives for plastics
   and thermosets
  *Tunable series of products adjustable to customer needs
  *Unique thermally stable form of inherently conductive
   polymers

Properties of the Eeonomer Additives:
  *Excellent thermal stability in ambient air (300 to 360C)
  *Tunable surface area, tribology, polarity, density, etc.
  *Improved dispersion and compatibility in plastics, thermo-
   sets, organic and aqueous solvents
  *Melt-processable into thermoplastics
  *Improved melt-flow behavior of Eeonomer blends
  *Improved mechanical properties of Eeonomer compounds
  *High conductivity of blends at moderate Eeonomer loading
  *Controllable and reproducible resistivity of Eeonomer
   loaded plastic compounds at static dissipative range

Eeonomer 200 Intrinsically Conductive Polypyrrole-Based Additive:
  Appearance: Black
  Bulk Conductivity: 30 S/cm
  Surface Resistivity: Down to 0.5 ohm/sq. (per customer spec.)
  Surface Area (BET-N2): 570 m2/g
  Particle Size*: avg. 40 nm
  Sieve Residue: >90% >600 mesh
  Water Content**: Less than 0.1%
  Ash Content: 0.01-0.04%
  Temperature Limits: Process up to at least 290C (560F)
  Solubility: not-soluble
  Chemical Nature: Neutral (per customer spec.)
  Apparent Density: 0.11 g/cm3

Eeonomer 500 Intrinsically Conductive Polyaniline-Based Additive:
  Appearance: Black
  Bulk Conductivity: 37 S/cm
  Surface Resistivity: down to 0.5 ohm/sq (per customer spec)
  Surface Area (BET-N2): 690 m2/g
  Particle Size*: avg. 40 nm
  Sieve Residue: >90% >600 mesh
  Water Content**: 1-2%
  Ash Content: 0.01-0.04%
  Temperature Limits: Process up to at least 290C (560F)
  Solubility: not-soluble
  Chemical Nature: Neutral (per customer spec.)
  Apparent Density: 0.11 g/cm3
  *Measured in polypropylene blend.
 **Recommend pre-drying at 125C for 1 hour prior to plastic
   hot-melt blending

Eeonyx Corp.: EEONOMER Conductive Additives (Continued):

**Eeonomer 760 Intrinsically Conductive Polyaniline-Based Additive
for Static Dissipative Applications:**
Appearance: Black
Bulk Conductivity: 5 to 10 S/cm
Surface Resistivity: down to 1.0 ohm/sq
Surface Area (BET-N2): 20 to 50 m2/g
Particle Size*: avg. 30 nm
Sieve Residue: >90% >600 mesh
Water Content**: less than 0.4%
Ash Content: 0.01-0.04%
Temperature Limits: Process up to at least 350C (662F)
Solubility: Partially soluble in organic solvents
Chemical Nature: Neutral
Apparent Density: 0.27 g/cm3

**Eeonomer 1350F Intrinsically Conductive Polyaniline-Based
Additive for Metals Corrosion Prevention Applications:**
Appearance: Black
Bulk Conductivity: 1.20 to 1.50 S/cm
Surface Resistivity: down to 1.0 ohm/sq
Surface Area (BET-N2): 6 to 12 m2/g
Particle Size*: avg. 40 nm
Sieve Residue: >90% >600 mesh
Water Content**: less than 1.0%
Ash Content: 0.01-0.04%
Temperature Limits: Process up to at least 250C
Solubility: not soluble
Chemical Nature: Moderately acidic
Apparent Density: 0.35 to 0.40 g/cm3

  *Measured in polypropylene blend
 **Recommend pre-drying at 125C for 1 hour prior to plastic
   hot-melt blending

Elm Grove Industries, Inc.: E/Z PURGE High Performance Purging Compound: 3 Levels of Purging Excellence:

E/Z Purge:
To be used as a quick, economical method to clean injection and blow molding machines and film and sheet extruders which are of a clean, smooth design with minimal changes of resin and routine, frequent cleaning, works well with PE and PP.

E/Z Purge Heavy Duty:
To be used where mold designs are more complicated, product runs are longer, color and resin changes more frequent, works well with PP, PE, Nylon and Polyester. Works well when mixed with scrap acrylic or any type of scrap resin for a more aggressive purge.

E/Z Purge Super:
To be used where mold designs are more complicated, product runs are shorter, color and resin changes more frequent, works well with more rigid materials, PVC, Nylon and Polycarbonate and especially with molds prone to produce burnt polymer. Works well when mixed with scrap acrylic or any type of scrap resin for a more aggressive purge.
All products are non-hazardous and water based

E/Z Purge is an ammonia-free, water-based product that removes and cleans unwanted color, material or burned polymer from extruders, injection or blow molding equipment. Purging is a critical part of routine maintenance but costly in terms of system downtime, wasted materials and labor costs. The efficient action of cleaning and purging with E/Z Purge enables you to make quick, complete changeovers from one color to another and from one resin type to another using your own scrap resin. E/Z Purge is safe for workers and the environment.

Features:
*No need to purchase more resin; mixes with your own scrap resin
*Ammonia-free
*Water-based
*Easy-feeding
*Odor free
*No residues
*Environmentally safe

Applications:
*Injection molding
*Blow molding
*Extrusion
*Co-extrusion

Materials:
*Polyester
*Nylon
*PVDC
*OPP/EVA
*LDPE
*HDPE
*PS, ABS, PVC, PP, PE

Lonza Group: ACRAWAX C Lubricant and Processing Aid:

N,N' Ethylene Bisstearamide
CAS No. 110-30-5

Form:	Beaded	Prilled
Acid Value:	8 Max	7 Max
Color, Gardner, 1963:	5 Max	4 Max
Melting point, C:	140-145	140-145
Neutralization Value	2.0	2.0
Screen Tests %:		
On 10 Mesh, %	10 Max	-----
On 40 Mesh, %	------	2 Max
On 100 Mesh, %	90 Min	-----
On 325 Mesh, %	------	90 Min
Typical Properties:		
Flash Point, C:	285	285
Mean Particle Size (microns)	570	130

Form:	Powdered	Atomized
Acid Value:	8 Max	6 Max
Color, Gardner, 1963	5 Max	5 Max
Melting Point, C:	140-145	140-145
Neutralization Value	2.0	2.0
Screen Tests %:		
On 100 Mesh, %	1 Max	---
On 325 Mesh, %	-----	1 Max
Typical Properties:		
Flash Point, C:	285	285
90% of Particles (Microns)	<140	<13
Mean Particle Size (Microns)	40	6

Acrawax C is also available in a 33% solid aqueous dispersion

Suggested Applications:
    Acrawax C Prilled or Beaded is an effective lubricant,
processing aid, slip additive and pigment dispersant aid for
most polymers including ABS, PVC, polypropylene, nylon, acetal,
polyethylene and thermoplastic polyester.
    Acrawax C Powdered and Atomized is traditionally used as a
lubricant and binder for cold compaction of powdered metal parts.

Mitsubishi Rayon Co., Ltd: METABLEN Processing Aids:

P-type: Processing aid
   P-570A   P-530A
   P-501A   P-700 (External lubricant type)
   P-550A   P-710 (High grade external lubricant type)
   P-551A
L-type: Acrylic external lubricant L-1000
H-type: Heat-distortion resistant modifier
   H-602 (Transparent type)

<----More excellent---
    AAA>AA>A>B>C

P-570A:
   Acceleration of fusion (gelation): AAA
   Vacuum forming processability: A-B
   Draw-down resistance: A-B
   Long run processability: A
   Processability in foaming extrusion: A-B
   Flexible purpose: A
   Appearance of injected products: A
   Melt viscosity (numbering from lower viscosity): 2

P-501A:
   Acceleration of fusion (gelation): A
   Vacuum forming processability: AA
   Draw-down resistance: AA
   Long run processability: A
   Processability in foaming extrusion: AA
   Flexible purpose: AA-A
   Appearance of injected products: AA
   Melt viscosity (numbering from lower viscosity): 4

P-550A:
   Acceleration of fusion (gelation): AA
   Vacuum forming processability: AA
   Draw-down resistance: AA
   Long run processability: A
   Processability in foaming extrusion: AA-A
   Flexible purpose: AA-A
   Appearance of injected products: AA
   Melt viscosity (numbering from lower viscosity): 4

P-551A:
   Acceleration of fusion (gelation): AAA
   Vacuum forming processability: AAA-AA
   Draw-down resistance: AAA-AA
   Long run processability: A
   Processability in foaming extrusion: AAA-AA
   Flexible purpose: AA
   Appearance of injected products: AAA-AA
   Melt viscosity (numbering from lower viscosity): 5

Mitsubishi Rayon Co., Ltd.: METABLEN Processing Aids (Continued):

P-530A:
   Acceleration of fusion (gelation): AAA-AA
   Vacuum forming processability: AAA
   Draw-down resistance: AAA
   Long run processability: A
   Processability in foaming extrusion: AAA
   Flexible purpose: AAA
   Appearance of injected products: AAA-AA
   Melt viscosity (numbering from lower viscosity): 6

P-700:
   Acceleration of fusion (gelation): B
   Vacuum forming processability: B
   Draw-down resistance: A
   Long run processability: AAA-AA
   Processability in foaming extrusion: B
   Flexible purpose: AAA-AA
   Appearance of injected products: A
   Melt viscosity (numbering from lower viscosity): 3

P-710:
   Acceleration of fusion (gelation): B
   Vacuum forming processability: B
   Draw-down resistance: A-B
   Long run processability: AAA-AA
   Processability in foaming extrusion: B
   Flexible purpose: AA
   Appearance of injected products: A
   Melt viscosity (numbering from lower viscosity): 1

L-1000:
   Draw-down resistance: B
   Long run processability: AAA
   Processability in foaming extrusion: B
   Flexible purpose: A
   Appearance of injected products: A-B
   Melt viscosity (numbering from lower viscosity): 2

H-602:
   Heat distortion resistance: AA
   Transparency: AAA
   Rigidity: AA
   Main applications: Sheets, profile extrusion

Tego Chemie Service GmbH: TEGO Glide & Flow Surface Control Additives:

Surface control additives are multi-functional, improving leveling, slip and scratch resistance, preventing craters and pigment floating.

Glide 100:
    System: W/S/UV
    Dosage: 0.1-1.0%
    Effect/Features: flow, slip, leveling

Glide ZG 400:
    System: W/S/UV
    Dosage: 0.05-1.0%
    Effect/Features: universal slip, prevention of Benard cells

Glide 406:
    System: W/S
    Dosage: 0.1-0.5%
    Effect/Features: slip, leveling

Glide 410:
    System: W/S/UV
    Dosage: 0.03-0.5%
    Effect/Features: maximum slip, anti-blocking

Glide 411:
    System: S
    Dosage: 0.3-1.0%
    Effect/Features: improves scratch resistance

Glide 415:
    System: S
    Dosage: 0.1-1.0%
    Effect/Features: high slip, flow, recoatability

Glide 420:
    System: S
    Dosage: 0.05-0.3%
    Effect/Features: deaeration, leveling, slip

Glide 435:
    System: W/UV
    Dosage: 0.05-1.0%
    Effect/Features: substrate wetting, flow, slip

Glide 440:
    System: W/UV
    Dosage: 0.05-1.0%
    Effect/Features: high slip, clarity, flow

Tego Chemie Service GmbH: TEGO Glide & Flow Surface Control Additives (Continued):

Glide 450:
    System: W/S
    Dosage: 0.03-0.2%
    Effect/Features: high slip, recoatability, flow

Glide A 115:
    System: S
    Dosage: 0.03-0.3%
    Effect/Features: improves scratch resistance, prevention of
        Benard cells, suitable for curtain coating

Glide B 1484:
    System: S
    Dosage: 0.05-1.0%
    Effect/Features: slip, flow, deaeration of epoxies

Flow ATF:
    System: S
    Dosage: 0.03-1.0%
    Effect/Features: anti-cratering, anti-blocking

Flow 300:
    System: S
    Dosage: 0.1-1.0%
    Effect/Features: leveling, flow, for clear systems, silicone-
        free

Flow 425:
    System: W/S/UV
    Dosage: 0.05-0.5%
    Effect/Features: leveling, flow, for clear systems

Flow ZFS 460:
    System: S/UV
    Dosage: 0.1-0.7%
    Effect/Features: deaeration, leveling, silicone-free

Tego Chemie Service GmbH: TEGO Rad Reactive Surface Control Additives:

Rad 2100:
   Dosage: 0.05-1.0%
   Effect/Features: flow, leveling

Rad 2200:
   Dosage: 0.05-1.0%
   Effect/Features: flow, substrate wetting, slip

Rad 2500:
   Dosage: 0.05-1.0%
   Effect/Features: slip, deaeration, release

Rad 2600:
   Dosage: 0.1-2.0%
   Effect/Features: release, anti-blocking, slip

Rad 2700:
   Dosage: 0.1-2.0%
   Effect/Features: extreme release, deaeration

Witco Corp.: Additives for Polyolefins: Lubrication/Processing Aids:

Calcium stearate, zinc stearate, potassium stearate and strontium stearate are used as lubricants to improve the flow characteristics of polyolefin resins. These stearates also act as stabilizers by performing acid scavenging in polymers manufactured using acidic polymerization catalysts.

Calcium Stearate F, EA, Kosher:
    FDA Sanctioned: Yes
    Acid Acceptor/Lubricant: EVA Modified PE
                             HDPE/LDPE/LLDPE
                             Polypropylene
                             UHMWPE

Potassium Stearate:
    FDA Sanctioned: Yes
    Lubricant: EVA Modified PE
               HDPE/LDPE/LLDPE
               Polyropylene
               UHMWPE

Zinc Stearate* ED--HS, Regular, Polymer Grade, Lubrazinc, W:
    FDA Sanctioned: Yes
    Acid Acceptor/Lubricant: EVA Modified PE
                             HDPE/LDPE/LLDPE
                             Polypropylene
                             UHMWPE
    Lubricant: Ionomer
    * Kosher grade available

Section XVII
Release Agents

Axel Plastics Research Laboratories Inc.: AXELMOLDWIZ:
Composite Molding: Mold Releases and Internal Lubricants:

Internal Mold Releases:
Polyester & Vinyl Ester:
MEK Peroxide Cured Resins & Gel Coats: INT-EQ6, INT-937,
INT-44-800
Clear Gel Coats: INT-389A, INT-22NR, INT-XL51
2,4 Pentanaedione Cured Resins: INT-XL51, INT-389A
Benzoyl Peroxide (BPO) Cured Resins: INT-PS125, INT-PUL24,
INT-54
Modar (MMA Modified Resins): INT-1866
T-Butyl Peroxide (TBPB) Cured Resins: INT-1938MCH, INT-PS-125,
INT-135PMC, INT-1208
T-Amyl Perbenzoate Cured Resins: INT-1938MCH, INT-PS-125,
INT-135PMC, INT-1208
Peroxyketal Cured Resins: INT-EQ6, INT-54, INT-44-800
Peroxydicarbonate Cured Resins: INT-PS125, INT-PS140
Dicumyl Peroxide Cured Resins: INT-PS125, INT-PS140
Dialkyl Peroxide Cured Resins: INT-1938MCH, INT-PS-125,
INT-135PMC, INT-1208

Epoxy:
Aliphatic Amine: INT-1846(N)
Aliphatic Amine-High Temp: INT-1850(HT)
Aromatic Amine-High Temp: INT-1824L
Anhydride (Clear): INT-1810
Anhydride: INT-1854
Anhydride-Amine or Imidazole Catalyzed: INT-1890M
Cycloaliphatic (PACM) Cure: INT-1840
Water-Based Epoxies: INT-EP545
UV Cured Epoxies, Epoxy/Acrylates: INT-1285C, INT-1210

Phenolics & Other Resins:
Clear Acrylic or MMA Modified Resins: INT-AM121, INT-389A,
INT-1866
Phenolic-Novolac Hexa Cured (Powdered): INT-8E-18S, INT-325PWD
Phenolic-Novolac Hexa Cured (Alcohol Solutions): INT-1838P,
INT-1858
Phenol-Resole (Water-Based): INT-4E-18CC, INT-1425PNP,
INT-1830PN, INT-EM972
Phenol-Resole (Alcohol-Based): INT-1312MS, INT-12
Phenol-Resole (Acid Cured): INT-1848PN

Axel Plastics Research Laboratories Inc.: AXEL MOLDWIZ: Polyurethane Mold Releases and Internal Lubricants:

Solvent-Based Mold Release Agents:
424/7NC:
   Elastomer/Flexible/Rigid/RIM

RTW850/NC:
   Rigid

RIMWIZ DU-21/NC:
   RIM

UFO-100:
   Seating

Film-Forming Barrier Release Agents:
FF1H:
   Rigid

Paste Release:
PasteWiz:
   Rigid

Water-Based Release Agents:
WB-411:
   Elastomer/Flexible/Rigid/RIM

WB-215A:
   Rigid

WB-640 Series:
   Rigid/RIM

H40-21U:
   Rigid

WPO-100:
   Elastomer

Internal Lubricants:
INT-20E:
   Elastomer

INT-XP420/2C:
   Flexible

INT-1230:
   Rigid/RIM

INT-1988A:
   Elastomer/Rigid/RIM

INT-PF160 (Phenolic Foam):
   Rigid

Axel Plastics Research Laboratories, Inc.: XTEND Semi-Permanent Mold Releases:

Solvent-Based:
800:
Used for: Open Molding/Class "A" Finishes/Cultured Marble
Application Technique: Wipe on
Application Temp: Ambient Temperature or Higher (<300F/148C)
Type of Molds: No Aluminum
Resin or Rubber: All except silicone & cyanate esters
Process Temperature: <400F/200C

802:
Used For: Like 800, but quicker evaporation
Application Technique: Wipe on or spray on
Application Temp: Ambient Temperature
Type of Molds: No aluminum
Resin or Rubber: All except silicone cynate esters
Process Temperature: <400F/200C

19RM:
Used For: Less Cosmetic than 800 Series/Excellent slip/
quick drying/curing
Application Technique/Application Temp: Wipe on/Ambient
Type of Molds: All mold surfaces
Resin or Rubber: All except silicone
Process Temperature: <400F/200C

19W:
Used For: Closed Molding-RTM/SMC/BMC
Application Technique: Wipe on Warm molds
Application Temp: 130-150F/55-85C
Type of Molds: All mold surfaces
Resin or Rubber: All except silicone
Process Temperature: <400/200C

19HG:
Used For: Closed Molding-RTM/Higher gloss; less buildup than
19W
Application Technique: Wipe on warm molds
Application Temp: 130-150F/55-65C
Type of Molds: All mold surfaces
Resin or Rubber: All except silicone
Process Temperature: <400F/200C

20RP:
Used For: For High Temperature Molding (up to 1000F/537C)
Especially for roto molding
Application Technique: Wipe on/Spray on
Application Temp: Ambient Temperature or higher
Type of Molds: All Mold Surfaces
Resin or Rubber: No phenolics
Process Temperature: Up to 1000F/537C

21:
Used For: For high temperature molding-No xylene (up to 600F)
Application Technique: Wipe on/spray on
Application Temp: Ambient temperature or higher
Type of Molds: All mold surfaces
Resin or Rubber: No phenolics
Process Temperature: Up to 600F/315C

**Axel Plastics Research Laboratories, Inc.: XTEND Semi-Permanent
Mold Releases (Continued):**

Water-Based:
W-120:
    Used For: Hot Pressing Applications/Hybrid Fluoropolymer
             Release/May be diluted
    Application Technique: Spray on
    Application Temp: Ambient or higher
    Type of Molds: All mold surfaces
    Resin or Rubber: Most
    Process Temperature: <400F/200C

W-3201:
    Used For: Hot Pressing Applications/Fluoropolymer Release
    Application Technique: Spray on warm-hot molds
    Application Temp: 180F/80C
    Type of Molds: All Mold Surfaces
    Resin or Rubber: All including silicone
    Process Temperature: <400F/200C

W-3202::
    Used For: Hot Pressing Applications/Fluoropolymer Release
    Application Technique/Temp: Spray on at ambient temperature
    Type of Molds: All mold surfaces
    Resin or Rubber: All including silicone
    Process Temperature: <400F/200C

W-3211:
    Used For: Hot pressing applications/Especially suitable for
             peroxide cured rubbers
    Application Technique: Spray on warm-hot molds
    Application Temp: 180F/80C
    Type of Molds: All mold surfaces
    Resin or Rubber: All including silicone
    Process Temperature: <400F/200C

W-4001:
    Used For: For high temperature molding (up to 1000F/537C)
    Application Technique: Spray on ambient or higher
    Application Temp: up to 200F/94C
    Type of Molds: All mold surfaces
    Resin or Rubber: All except silicone
    Process Temperature: Up to 1000F/537C

W-7200:
    Used For: Epoxy & epoxy pre-preg
    Application Technique: Spray on ambient or higher
    Application Temp: Up to 200F/94C
    Type of Molds: Steel
    Resin or Rubber: All except silicone
    Process Temperature: <400F/200C

Slide Products, Inc.: Mold Releases:

Light Duty Releases:

ECONOMIST:
Silicone release   Non-staining
FDA Approved Lubricant   Odorless
Approved for food machinery(++)
Maximum operating temperature: 600F

Paintable:
Light-duty biodegradable release No spalling & no fish eyes
Paint over it   Perfect finish
Plate over it   Odorless
Hot-stamp over it
Maximum operating temperature: 650F
Recognized by Underwriters Laboratories

LMR Lecithin:
Non-silicone paintable   A direct food additive(+)
Light-duty natural release   Biodegradable & odorless
FDA Approved Lubricant
Maximum Operating Temperature: 500F

Medium-Duty Mold Releases:

Quick Silicone:
All-temperature silicone release No chlorinated solvents
FDA Approved Lubricant   Odorless
Fast-drying on cold and hot molds
Operating temperatures: 35F-600F

Quick Paintable:
All-temperature paintable release No chlorinated solvents
Fast-drying on cold and hot molds Biodegradable & odorless
Operating temperatures: 35F-650F
Recognized by Underwriters Laboratories

Quick Lecithin:
Non-silicone paintable   Fast-drying on cold & hot molds
All-temperature lecithin release No chlorinated solvents
FDA Approved lubricant   Biodegradable & odorless
Operating temperatures: 35F-500F

Slide Products, Inc.: Mold Releases (Continued):

Heavy Duty Mold Releases:

Silicone:
FDA Approved lubricant   Heat stable & non-toxic
No ozone depleting chemicals Approved for food machinery (++)
Faster molding & more production   Lubricates
Eliminates rejects   Non-staining
Maximum operating temperature: 600F

Polycarbonate 41412N:
Paintable release   No crazing or blemishing
No ozone depleting chemicals Biodegradable & odorless
Maximum operating temperature: 650F
Recognized by Underwriters Laboratories

Universal:
Non-silicone paintable   Use where parts must be painted,
No ozone depleting chemicals   hot-stamped or metallized
FDA Approved lubricant   Biodegradable & odorless
Maximum operating temperature: 600F
Recognized by Underwriters Laboratories

Specialty Mold Releases:

Dry Film Lube (Fluorocarbon):
Formulation for deep draw molds No ozone depleting chemicals
Excellent for most thermoplastics: ABC, acetyl, nylon, vinyl,
PVC   Also good on phenols & urethane
Contains Krytox, PTFE
Maximum operating temperature: 500F

Electronic:
Non-silicone paintable   For plastic electronic parts
Maximum operating temperature: 550F

Hi-Temp 1000:
Non-silicone paintable   For temperatures up to 1000F
No ozone depleting chemicals
Maximum operating temperature: 1000F

Mold Saver:
Non-silicone paintable   Neutralizes corrosive vapors
No ozone depleting chemicals Eliminates mold build-up
Prevents deposit build-up
Maximum operating temperature: 550F

Slide Products Inc.: Mold Releases (Continued):

Specialty Mold Releases (Continued):

Pure Eze:
    Non-silicone paintable   An excellent all-purpose release
    Neutral white-oil-based release++++   No lecithins
    Won't turn color or turn rancid   USDA rated H1
Maximum operating temperature: 600F
Recognized by Underwriters Laboratories

Water Soluble:
    Non-silicone paintable   No ozone depleting chemicals
    Permits ultrasonic welding of parts   No removal necessary
    Does not contain water
Maximum operating temperature: 450F

Zinc Stearate:
    Non-silicone paintable   Lubricant powder
    For polypropylenes, polysulfones and rubber molding
Maximum operating temperature: 600F

Water-Based Releases: E/S Environmentally Safe:

E/S Environmentally Safe Silicone:
    CFC-free & chlorine-free   No ozone depleting chemicals
    No VOC's   FDA and USDA Approved (+++)
    Biodegradable   Non-flammable
    Colorless and non-staining
Maximum operating temperature: 600F

E/S Environmentally Safe Paintable:
    CFC-free & chlorine-free   No ozone depleting chemicals
    No VOC's   Biodegradable
    Non-flammable   Colorless, odorless and non-staining
Maximum operating temperature: 650F

E/S Environmentally Safe Paintable Lecithin:
    Non-silicone paintable   CFC-free & chlorine-free
    No ozone depleting chemicals No VOC's & Non-flammable
    FDA and USDA Approved(+)   Biodegradable
    Colorless, odorless and non-staining
Maximum operating temperature: 600F

Slide Products, Inc.: Mold Releases (Continued):

Internal Releases:

Zinc Stearate Inter Lube Internal Mold Release:
    Lubricant powder

Thermoset Mold Releases:

Urethane Mold Release:
    For rigid, semi-rigid and flexible urethane foams
    No ozone depleting chemicals
    Non-toxic & no chlorinated solvent
    Made especially for polyurethanes
    Non-paintable silicone
    Non-marking
Maximum operating temperature: 600F

Dura Kote:
    Semi-permanent for thermosets   Will not discolor
    No ozone depleting chemicals   Extremely thin film
    Use for urethane & epoxy molding Paintable
    Completely dry
    Excellent rotational mold release
Maximum operating temperature: 600F

EPOXEASE Mold Release:
    Non-silicone for thermosets   No ozone depleting chemicals
    Synthetic wax-based product
    For injection molding, encapsulating, potting
    For epoxy, polyester and phenolic molding
Maximum operating temperature: 450F

Thermoset Mold Release:
    Made with pure wax
    No ozone depleting chemicals
    Contains no silicones or oils
Maximum operating temperature: 550F

Stoner Inc.: Mold Releases:

Light Duty Silicone Mold Releases:
   *Food grade
   *Prevents mold build-up

Silicone Mold Release (Item #E206):
   *Food grade
   *Versatile, high performance

Heavy Duty Silicone Mold Releases:
   *E208 is food grade
   *For extra stubborn parts

Urethane Mold Release (Item #E236):
   *Best for urethanes
   *Non-flammable

Rocket Release (Item #E302):
   *Paintable
   *Food grade

Release & Paint (Item #E313):
   *Paintable
   *UL recognized

Zero Stick (Item #E342):
   *Paintable
   *Food grade

Dry Film Mold Release (Item #E408):
   *Paintable
   *Highly versatile

Mold Release and Ejector Pin Lube (Item #E436):
   *Paintable
   *Food grade

TPO Release (Item #E464):
   *Paintable
   *For thermoplastic olefins

Zinc Stearate Mold Release (Item #E474):
   *Paintable
   *Food grade

Thermoset Mold Releases:
   *Paintable
   *No silicones or oils

Witco Corp.: Additives for Polyolefins: Mold Release Agents:

Witco internal mold release agents are an alternative to spray-on release agents. They provide a consistent, predictable amount of release without the drawbacks of fluorocarbon or silicon spray-on release agents. Kemamide E Ultra fatty amide is recommended for polyethylenes. Atmul 84, Atmos 150, and Dimul S additives provide a combination of mold release and antistat for polypropylene. Kemamide AS-974, AS-974/1, and AS-989 fatty amines provide a combination mold release and antistat for polyethylenes.

Kemamide E Ultra:
    FDA Sanctioned: Yes
    Antiblock/Slip: EVA modified PE
                    UHMWPE
    Antiblock/Mold Release/Slip: HDPE/LDPE/LLDPE
                                 Polypropylene

Atmul 84:*
    FDA Sanctioned: Yes
    Foaming Aid: Foamed PE
    Antifog: LDPE, LLDPE
    Antifog/Mold Release: Polypropylene

Atmos 150:*
    FDA Sanctioned: Yes
    Foaming Aid: Foamed PE
    Antifog/Antistat: LDPE, LLDPE
    Antifog/Antistat/Mold Release: Polypropylene

Dimul S:*
    FDA Sanctioned: Yes
    Antistat: EVA Modified PE
    Antifog/Antistat: LDPE, LLDPE
    Antifog/Mold Release: Polypropylene
    * Kosher Grade Available

Kemamine AS-974/Kemamine AS-974/1:
    FDA Sanctioned: Yes
    Antifog/Antistat: EVA Modified PE
                      LDPE, LLDPE
                      Polypropylene
    Antifog/Antistat/Mold Release: HDPE

Kemamine AS-989:
    FDA Sanctioned: Yes
    Antifog/Antistat: EVA Modified PE
                      LDPE/LLDPE
                      Polypropylene
    Antifog/Antistat/Mold Release: HDPE

Section XVIII

Silanes, Titanates
and Zirconates

Genessee Polymers Corp.: Silanes:

EXP-49 Dimethyldiethoxy Silane:

EXP-49 Dimethyldiethoxy Silane is an intermediate useful for blocking hydroxyl and amino groups in organic synthesis reactions. This silating step allows subsequent reactions to be carried out which would be adversely affected by the presence of active hydrogen in the hydroxyl or amine groups. Following the reaction step, hydroxyl or amine groups blocked with EXP-49 may be recovered by a hydrolysis procedure. EXP-49 is also used for preparing hydrophobic and release materials as well as enhancing flow of powders.

Typical Properties:

    Appearance: Clear, Colorless Liquid
    Boiling Range: 237-239F
    Flash Point (P.M.C.C.): 29F
    Specific Gravity: 0.84
    Wt./Gallon @ 68F: 7.0 lbs.
                @ 77F: 6.9 lbs.

EXP-51 Trimethylethoxy Silane:

EXP-51 Trimethylethoxy Silane is an intermediate useful for blocking hydroxyl or amino groups to perform reactions on multi-functional organic compounds or polymers. It is also useful for deactivating glass surfaces used in gas chromatographic applications.

Typical Properties:

    Appearance: Clear, Colorless Liquid
    Wt./Gallon @ 68F: 6.3
    Flash Point (P.M.C.C.): 23F
    Specific Gravity: 0.75
    Boiling Range: 167-169F
    Vapor Pressure @ 77F: 111 mm Hg

Kenrich Petrochemicals, Inc.: Chelate Titanate Coupling Agents: A,B Ethylene Chelate Type:

KR 212:
   di(dioctyl)phosphato, ethylene titanate
   Liquid
   Color: Descriptive: Pale Orange-Red
       Gardner: 4
   Specific Gravity @ 16C: 0.98
   Solids %: 75+
   Viscosity @ 25C, cps: 300
   Flash Point F (TCC): 70
   Initial Boiling Point F: 185
   pH (Saturated Solution): 4.8

KR 238S:
   di(dioctyl)pyrophosphato ethylene titanate
   Liquid
   Color: Descriptive: Pale Orange-Red
       Gardner: 4
   Specific Gravity @ 16C: 1.08
   Solids %: 78+
   Viscosity @ 25C, cps: 800
   Flash Point F (TCC): 70
   Initial Boiling Point F: 160
   pH (saturated Solution): 3

KR 262ES:
   di(butyl, methyl)pyrophospato, ethylene titanate
   Physical Form: Liquid
   Color: Descriptive: Yellow-Tan
       Gardner: 1
   Specific Gravity @ 16C: 1.17
   Solids %: 82+
   Viscosity @ 25C, cps: 100
   Flash Point F (TCC): 85
   Initial Boiling Point F: 170
   pH (saturated Solution): 2

Kenrich Petrochemicals, Inc.: Chelate Titanate Coupling Agent
Oxyacetate Chelate Type:

KR 134S:
di(cumyl)phenyl oxoethylene titanate
Liquid
Color: Descriptive: Red
      Gardner: >18
Specific Gravity @ 16C: 1.14
Solids %: 95+
Viscosity @ 25C, cps: 8000
Flash Point F (TCC): 195
Initial Boiling Point F: 180
pH (Saturated Solution): 5

KR 138S:
di(dioctyl)pyrophosphate oxoethylene titanate
Liquid
Color: Descriptive: Pale Yellow
      Gardner: 4
Specific Gravity @ 16C: 1.12
Solids %: 99+
Viscosity @ 25C, cps: 1000
Flash Point F (TCC): 100
Initial Boiling Point F: 160
pH (Saturated Solution): 3

KR 133DS:
dimethacryl. oxoethylene titanate
Liquid
Color: Descriptive: Amber
      Gardner: 2
Specific Gravity @ 16C: 1.06
Solids %: 45+
Viscosity @ 25C, cps: 100
Flash Point F (TCC): 150
Initial Boiling Point F: 272
pH (Saturated Solution): 2.5

KR 158FS:
di(butyl, methyl)pyrophosphato, oxoethylene di(dioctyl)
phosphito titanate
Liquid
Color: Descriptive: Yellow-Tan
      Gardner: 5
Specific Gravity @ 16C: 1.16
Solids %: 80+
Viscosity @ 25C, cps: <200
Flash Point F (TCC): 110
Initial Boiling Point F: 170
pH (Saturated Solution): 3

Kenrich Petrochemicals Inc.: Coordinate Titanate and Zirconate Coupling Agents:

KR 418:
    tetraisopropyl di(dioctyl)phosphito titanate
    Liquid
    Color: Yellow/Gardner: 4
    Specific Gravity @ 16C: 0.96
    Solids in IPA Solvent %: 98+
    Viscosity @ 25C, cps: 15
    Flash Point F (TCC): 130
    Initial Boiling Point F: 160
    pH (Saturated Solution): 6

KR 46B:
    tetraoctyl di(ditridecyl)phosphito titanate
    Liquid
    Color: Pale Yellow/Gardner: 2
    Specific Gravity @ 16C: 0.92
    Solids in IPA Solvent %: 98+
    Viscosity @ 25C, cps: 50
    Flash Point F (TCC): 180
    Initial Boiling Point F: 160
    pH (Saturated Solution): 6

KR 55:
    tetra (2,2-diallyoxymethyl)butyl, di(ditridecyl)phosphito titanate
    Liquid
    Color: Yellow/Gardner: 2
    Specific Gravity @ 16C: 0.97
    Solids in IPA Solvent %: 98+
    Viscosity @ 25C, cps: 50
    Flash Point F (TCC): 180
    Initial Boiling Point F: 120
    pH (Saturated Solution): 5

KZ 55:
    tetra (2,2 diallyloxymethyl)butyl, di(ditridecyl)phosphito zirconate
    Liquid
    Color: Light Brown/Gardner: 4
    Specific Gravity @ 16C: 1.00
    Solids in IPA Solvent %: 90+
    Viscosity @ 25C, cps: 100
    Flash Point F (TCC): >200
    Initial Boiling Point F: 380
    pH (Saturated Solution): 5.7

Kenrich Petrochemicals Inc.: Cycloheteroatom Titanate and Zirconate Coupling Agents:

KR OPPR:
Cyclo(dioctyl)pyrophosphato dioctyl titanate
Liquid
Color: Yellowish Brown/Gardner: 11
Specific Gravity @ 16C: 1.06
Solids in IPA Solvent %: 66
Viscosity @ 25C, cps: 110
Flash Point F (TCC): 160
Initial Boiling Point F: 280
pH (Saturated Solution): 6

KR OPP2:
Dicyclo(dioctyl)pyrophosphato, titanate
Liquid
Color: Yellowish Brown/Gardner: 13
Specific Gravity @ 16C: 1.13
Solids in IPA Solvent %: 66
Viscosity @ 25C, cps: 7000
Flash Point F (TCC): 140
Initial Boiling Point F: 210
pH (Saturated Solution): 5

KZ OPPR:
Cyclo(dioctyl)pyrophosphato dioctyl zirconate
Liquid
Color: Pale Yellow/Gardner: 2
Specific Gravity @ 16C: 1.12
Solids in IPA Solvent %: 90
Viscosity @ 25C, cps: 1000
Flash Point F (TCC): 150
Initial Boiling Point F: N/E
pH (Saturated Solution): 5

KZ TPP:
Cyclo[dineopentyl(diallyl)] pyrophosphato dineopentyl-
(diallyl) zirconate
Liquid
Color: Yellowish Brown/Gardner: 12
Specific Gravity @ 16C: 1.18
Solids in IPA Solvent %: 95
Viscosity @ 25C, cps: 3280
Flash Point F (TCC): >200
Initial Boiling Point F: 170
pH (Saturated Solution): 5.7

Kenrich Petrochemicals Inc.: Monoalkoxy Titanate Coupling Agents:

KR TTS:
isopropyl triisostearoyl titanate
Liquid
Color: Descriptive: Transparent Reddish Brown
      Gardner: 18
Specific Gravity @ 16C: 0.95
Solids in IPA Solvent %: 95+
Viscosity @ 25C, cps: 125
Flash Point F (TCC): 200+
Initial Boiling Point F: 300
pH (Saturated Solution): 5.5

KR 7:
isopropyl dimethacryl isoistearoyl titanate
Liquid
Color: Descriptive: Dark Red Brown
      Gardner: 14
Specific Gravity @ 16C: 1.02
Solids in IPA Solvent %: 95+
Viscosity @ 25C, cps: 220
Flash Point F (TCC): 110
Initial Boiling Point F: 250
pH (Saturated Solution): 5

KR 9S:
isopropyl tri(dodecyl)benzenesulfonyl titanate
Liquid
Color: Descriptive: Transparent Reddish Brown
      Gardner: 10
Specific Gravity @ 16C: 1.08
Solids in IPA Solvent %: 88+
Viscosity @ 25C, cps: 8000
Flash Point F (TCC): 97
Initial Boiling Point F: 130
pH (Saturated Solution): 2

KR 12:
isopropyl tri(dioctyl)phosphato titanate
Liquid
Color: Descriptive: Transparent to Translucent Off-White
      Gardner: 8
Specific Gravity @ 16C: 1.04
Solids in IPA Solvent %: 95+
Viscosity @ 25C, cps: 1500
Flash Point F (TCC): 150
Initial Boiling Point F: 170
pH (Saturated Solution): 4.5

Kenrich Petrochemicals Inc.: Monoalkoxy Titanate Coupling Agents (Continued):

KR 26S:
    isopropyl (4-amino)benzenesulfonyl di(dodecyl)benzenesulfonyl
titanate
    Liquid
    Color: Descriptive: Grey/Gardner: >18
    Specific Gravity @ 16C: 1.12
    Solids in IPA Solvent %: 95+
    Viscosity @ 25C, cps %: 30,000
    Flash Point F (TCC): 75
    Initial Boiling Point F: 250
    pH (Saturated Solution): 6.5

KR 33DS:
    Liquid
    alkoxy trimethacryl titanate
    Color: Descriptive: Tan to Red Brown/Gardner: 13
    Specific Gravity @ 16C: 1.11
    Solids in IPA Solvent %: 78+
    Viscosity @ 25C, cps: 100
    Flash Point F (TCC): 140
    Initial Boiling Point F: 170
    pH (Saturated Solution): 3.5

KR 38S:
    Liquid
    isopropyl tri(dioctyl)pyrophosphato titanate
    Color: Descriptive: Yellow to Amber/Gardner: 7
    Specific Gravity @ 16C: 1.09
    Solids in IPA Solvent %: 99+
    Viscosity @ 25C, cps: 1500
    Flash Point F (TCC): 100
    Initial Boiling Point F: 170
    pH (Saturated Solution): 2

KR 39DS:
    alkoxy triacryl titanate
    Liquid
    Color: Descriptive: Tan to Red Brown Liquid/Gardner: 18
    Specific Gravity @ 16C: 1.12
    Solids in IPA Solvent %: 49+
    Viscosity @ 25C, cps: 100
    Flash Point F (TCC): 180
    Initial Boiling Point F: 220
    pH (Saturated Solution): 3

KR 44:
    isopropyl tri(N-ethylenediamino)ethyl titanate
    Liquid
    Color: Descriptive: Yellow Brown/Gardner: 10
    Specific Gravity @ 16C: 1.2
    Solids in IPA Solvent %: 95+
    Viscosity @ 25C, cps: 10,000
    Flash Point F (TCC): 130
    Initial Boiling Point F: 180
    pH (Saturated Solution): 10

Kenrich Petrochemicals Inc.: Neoalkoxy Titanate Coupling Agents:

LICA 01:
   neopentyl(diallyl)oxy, trineodecanoyl titanate
   Liquid
   Color: Brownish Orange/Gardner: 10
   Specific Gravity @ 16C: 1.02
   Solids in IPA Solvent %: 95
   Viscosity @ 25C, cps: 850
   Flash Point F (TCC): 160
   Initial Boiling Point F: 320
   pH (Saturated Solution): 5

LICA 09:
   neopentyl(diallyl)oxy, tri(dodecyl)benzene-sulfonyl titanate
   Liquid
   Color: Greenish Brown/Gardner: 9
   Specific Gravity @ 16C: 1.04
   Solids in IPA Solvent %: 90
   Viscosity @ 25C, cps: 2000
   Flash Point F (TCC): 180
   Initial Boiling Point F: 170
   pH (Saturated Solution): 2

LICA 12:
   neopentyl(diallyl)oxy, tri(dioctyl)phosphato titanate
   Liquid
   Color: Brownish Orange/Gardner: 5
   Specific Gravity @ 16C: 1.03
   Solids in IPA Solvent %: 95
   Viscosity @ 25C, cps: 1800
   Flash Point F (TCC): 125
   Initial Boiling Point F: 160
   pH (Saturated Solution): 5.5

LICA 38:
   neopentyl(diallyl)oxy, tri(dioctyl)pyro-phosphato titanate
   Liquid
   Color: Greenish Brown/Gardner: 15
   Specific Gravity @ 16C: 1.13
   Solids in IPA Solvent %: 95
   Viscosity @ 25C, cps: <5000
   Flash Point F (TCC): 140
   Initial Boiling Point F: 160
   pH (Saturated Solution): 3.5

Kenrich Petrochemicals Inc.: Neoalkoxy Titanate Coupling Agents: (Continued):

LICA 44:
    neopentyl(diallyl)oxy, tri(N-ethylenediamino) ethyl titanate
    Liquid
    Color: Brownish Orange/Gardner: 6
    Specific Gravity @ 16C: 1.17
    Solids in IPA Solvent %: 95
    Viscosity @ 25C, cps: 10,000
    Flash Point F (TCC): 200
    Initial Boiling Point F: 250
    pH (Saturated Solution): 11

LICA 97:
    neopentyl(diallyl)oxy, tri(m-amino)phenyl titanate
    Liquid
    Color: Brown/Gardner: 18
    Specific Gravity @ 16C: 1.17
    Solids in IPA Solvent %: 56
    Viscosity @ 25C, cps: 10,000
    Flash Point F (TCC): 160
    Initial Boiling Point F: 180
    pH (Saturated Solution): 6

LICA 99:
    neopentyl(diallyl)oxy, trihydroxy caproyl titanate
    Liquid
    Color: Tan
    Specific Gravity @ 16C: 1.03

Kenrich Petrochemicals Inc.: Neoalkoxy Zirconate Coupling Agents:

NZ 01:
    neopentyl(diallyl)oxy, trineodecanoyl zirconate
    Liquid
    Color: Amber/Gardner: 11
    Specific Gravity @ 16C: 1.06
    Solids in IPA Solvent %: 95
    Viscosity @ 25C, cps: 400
    Flash Point F (TCC): 195
    Initial Boiling Point F: 188
    pH (Saturated Solution): 8

NZ 09:
    neopentyl(diallyl)oxy, tri(dodecyl)benzene-sulfonyl zirconate
    Liquid
    Color: Brown/Gardner: 18
    Specific Gravity @ 16C: 1.09
    Solids in IPA Solvent %: 95
    Viscosity @ 25C, cps: 15,000
    Flash Point F (TCC): 150
    Initial Boiling Point F: 300
    pH (Saturated Solution): 4

NZ 12:
    neopentyl(diallyl)oxy, tri(dioctyl)pyro-phosphato zirconate
    Liquid
    Color: Orange/Red/Gardner: 4
    Specific Gravity @ 16C: 1.06
    Solids in IPA Solvent %: 95
    Viscosity @ 25C, cps: 160
    Flash Point F (TCC): 170
    Initial Boiling Point F: 220
    pH (Saturated Solution): 6

NZ 38:
    neopentyl(diallyl)oxy, tri(dioctyl)pyro-phosphato zirconate
    Liquid
    Color: Reddish/Gardner: 12
    Specific Gravity @ 16C: 1.10
    Solids in IPA Solvent %: 95
    Viscosity @ 25C, cps: 3000
    Flash Point F (TCC): 170
    Initial Boiling Point F: 345
    pH (Saturated Solution): 6

NZ 44:
    neopentyl(diallyl)oxy, tri(N-ethylenediamino) ethyl zirconate
    Liquid
    Color: Yellow/Orange/Gardner: 13
    Specific Gravity @ 16C: 1.17
    Solids in IPA Solvent %: 95
    Viscosity @ 25C, cps: 5000
    Flash Point F (TCC): >220
    Initial Boiling Point F: 300

Kenrich Petrochemicals Inc.: Neoalkoxy Zirconate Coupling Agents (Continued):

NZ 97:
neopentyl(diallyl)oxy, tri(m-amino)phenyl zirconate
Liquid
Color: Brown/Gardner: >18
Specific Gravity @ 16C: 1.20
Solids in IPA Solvent %: 67
Viscosity @ 25C, cps: 8000
Flash Point F (TCC): 190
Initial Boiling Point F: 300
pH (Saturated Solution): 6

NZ 33:
neopentyl(diallyl)oxy, trimethacryl zirconate
Liquid
Color: Yellow/Orange/Gardner: <5
Specific Gravity @ 16C: 1.09
Solids in IPA Solvent %: 95+
Viscosity @ 25C, cps: <100
Flash Point F (TCC): 150
Initial Boiling Point F: >300
pH (Saturated Solution):4

NZ 39:
neopentyl(diallyl)oxy, triacryl zirconate
Liquid
Color: Yellow/Orange/Gardner: <5
Specific Gravity @ 16C: 1.07
Solids in IPA Solvent %: 95+
Viscosity @ 25C, cps: <100
Flash Point F (TCC): <200
Initial Boiling Point F: >300
pH (Saturated Solution): 4

NZ 37:
dineopentyl(diallyl)oxy diparamino benzoyl zirconate
Liquid
Color: Dark Brown/Gardner: >18
Specific Gravity @ 16C: 1.12
Solids in IPA Solvent %: 46
Viscosity @ 25C, cps: 100
Flash Point F (TCC): 200
Initial Boiling Point F: 300
pH (Saturated Solution): 6

NZ 66A:
dineopentyl(diallyl)oxy, di(3-mercapto) propionic zirconate
Liquid
Color: Lemon Yellow/Gardner: 4
Specific Gravity @ 16C: 1.18
Solids in IPA Solvent %: 56
Viscosity @ 25C, cps: 250
Flash Point F (TCC): 200
Initial Boiling Point F: 230
pH (Saturated Solution): 6

Kenrich Petrochemicals Inc.: Quat Titanate and Zirconate Coupling Agents:

KR 138D:
  2-(N,N-dimethylamino)isobutanol adduct of KR 138S
  Liquid
  Color: Tan/Brown/Gardner: 2
  Specific Gravity @ 16C: 1.06
  Solids %: 94+
  Viscosity @ 25C, cps: <1000
  Flash Point F (TCC): 95
  Initial Boiling Point F: 180
  pH (Saturated Solution): 8
KR 158D:
  2-(N,N-dimethylamino)isobutanol adduct of KR 158
  Liquid
  Color: Pale Yellow/Gardner: 2
  Specific Gravity @ 16C: 1.08
  Solids %: 92
  Viscosity @ 25C, cps: 400
  Flash Point F (TCC): 85
  Initial Boiling Point F: 162
  pH (Saturated Solution): 7.2
KR 238T:
  triethylamine adduct of KR 238S
  Liquid
  Color: Pale Yellow to Orange/Red/Gardner: 10
  Specific Gravity @ 16C: 1.02
  Solids %: 95+
  Viscosity @ 25C, cps: <200
  Flash Point F (TCC): 100
  Initial Boiling Point F: 170
  pH (Saturated Solution): 7
KR 238M:
  methacrylate functional amine adduct of KR 238S
  Liquid
  Color: Yellow to Amber/Gardner: 10
  Specific Gravity @ 16C: 1.04
  Solids %: 94+
  Viscosity @ 25C, cps: <400
  Flash Point F (TCC): 95
  Initial Boiling Point F: 170
  pH (Saturated Solution): 5.5
KR 238A:
  acrylate functional amine adduct of KR 238S
  Liquid
  Color: Yellow/Gardner: 9
  Specific Gravity @ 16C: 1.03
  Solids %: 95+
  Viscosity @ 25C, cps: <300
  Flash Point F (TCC): 95
  Initial Boiling Point F: 170
  pH (Saturated Solution): 8

Kenrich Petrochemicals Inc.: Quat Titanate and Zirconate Coupling Agents (Continued):

KR 238J:
methacrylamide functional amine adduct of KR 238S
Liquid
Color: Amber/Red/Gardner 13
Specific Gravity @ 16C: 1.06
Solids %: 88+
Viscosity @ 25C, cps: <1200
Flash Point F (TCC): 85
Initial Boiling Point F: 220
pH (Saturated Solution): 7

KR 262A:
acrylate functional amine adduct of KR 238S
Liquid
Color: Pale Yellow/Gardner: 9
Specific Gravity @ 16C: 1.11
Solids %: 88+
Viscosity @ 25C, cps: <5000
Flash Point F (TCC): 95
Initial Boiling Point F: 170
pH (Saturated Solution): 6.5

LICA 38J:
methacrylamide functional amine adduct of LICA 38
Liquid
Color: Yellowish-Red/Gardner: 17
Specific Gravity @ 16C: 1.09
Solids %: 95+
Viscosity @ 25C, cps: 5100
Flash Point F (TCC): 160
Initial Boiling Point F: 220
pH (Saturated Solution): 7.5

NZ 38J:
methacrylamide functional amine adduct of a LZ 38 analog
Liquid
Color: Amber Orange/Gardner: 12
Specific Gravity @ 16C: 1.02
Solids %: 85+
Viscosity @ 25C, cps: 1200
Flash Point F (TCC): 90
Initial Boiling Point F: 220
pH (Saturated Solution): 9

KZ TPPJ:
Cycloneopentyl, cyclo(dimethylaminoethyl) pyrophosphato
zirconate, di mesyl salt
Liquid
Color: Tan
Specific Gravity @ 16C: 1.08

Section XIX
Slip and Anti-Blocking Agents

**Akzo Nobel Chemicals Inc.: ARMOSLIP Slip and Antiblocking
  Additives:**

Armoslip 18:
  Physical Form: Flakes/Pellets
  Amide (% min): 90.0
  Melting Point (C): 99
  Specific Gravity: 0.834
  Flash Point COC (C): >225
  TGA Rapid Weight Loss Temperature (C): 235

Armoslip CP:
  Physical Form: Flakes/Pellets
  Amide (% min): 97.0
  Melting Point (C): 71
  Specific Gravity: 0.845
  Flash Point COC (C): >210
  TGA Rapid Weight Loss Temperature (C): 254

Armoslip EXP:
  Physical Form: Flakes/Pellets
  Amide (% min): 97.0
  Melting Point (C): 80
  Specific Gravity: 0.837
  Flash Point COC (C): >230
  TGA Rapid Weight Loss Temperature (C): 243

Armoslip HT:
  Physical Form: Flakes/Pellets
  Amide (% min): 90.0
  Melting Point (C): 98
  Specific Gravity: 0.830
  Flash Point COC (C): >225
  TGA Rapid Weight Loss Temperature (C): 264

Armoslip OPA:
  Physical Form: Flakes
  Amide (% min): 94.0
  Melting Point (C): 69
  Specific Gravity: 0.81
  Flash Point COC (C): 260
  TGA Rapid Weight Loss Temperature (C): 310

Armoslip SSA:
  Physical Form: Flakes
  Amide (% min): 90.0
  Melting Point (C): 91
  Specific Gravity: 0.804
  Flash Point COC (C): 246
  TGA Rapid Weight Loss Temperature (C): 295

Chemax, Inc.: CHEMSTAT Slip Agents:

HTSA #1A:
   Use Level % by Weight: PP: 1.0-3.0
   Chemical Composition: Oleyl Palmitamide
   Form: Bead
   FDA: Yes

HTSA #3B:
   Use Level % by Weight: PVC: 0.50-5.0//Nylon: 0.50-0.80
   Chemical Composition: Stearyl Erucamide
   Form: Bead
   FDA: Yes

HTSA #18:
   Use Level % by Weight: PE: 0.50-1.5//PP: 0.50-1.5
   Chemical Composition: Oleamide
   Form: Bead
   FDA: Yes

HTSA #18S:
   Use Level % by Weight: PE: 0.20-0.30//PP: 0.20-0.30
   Chemical Composition: Stearamide
   Form: Bead
   FDA: Yes

HTSA #22:
   Use Level % by Weight: PE: 0.05-1.0//PP: 0.05-1.0
   Chemical Composition: Erucamide
   Form: Bead
   FDA: Yes

HTSA #54:
   Use Level % by Weight: PVC: 0.80-1.2/PC: 0.80-1.2
   Chemical Composition: Polyol ester
   Form: Bead
   FDA: Yes

Degussa AG: SIPERNAT 44 Antiblocking Agent for Foils; Especially, Polyethylene Foil:

Sipernat 44 is a synthetic sodium aluminum silicate which corresponds by approximation to the formula $Na_2O-Al_2O_3-2SiO_2-4H_2O$ With regard to the chemical composition and the physical properties, this product differs basically from the synthetic silicas (silica gels and precipitated silicas) and natural products (diatomaceous earth) which have been used up to now as antiblocking agents.

On account of the method of production, Sipernat 44 is free from impurities which could lead to discolouration. Even a masterbatch of high concentration shows only a white brightening. There is no yellow or grey discoloration, a phenomenom which is often observed when using other materials. Furthermore, the method of synthetic production and the raw materials used guarantee that a constant quality of Sipernat 44 is delivered.

The values for "particle size" usually indicated for antiblocking agents depend highly on the corresponding test methods applied. These methods are often not comparable to each other.

Incorporation:
In comparison with other antiblocking agents, Sipernat 44 is relatively readily dispersible. Nevertheless, it is advisable to incorporate Sipernat 44, too, according to one of the usual methods (in most cases kneader) to a masterbatch (concentrate).

Necessary Quantity:
According to the thickness of the foil and the density/stiffness of the polymer, it is sufficient to add only approx. 0.1% Sipernat 44 in order to achieve a good effect of antiblocking. This low quantity almost does not influence the transparency, lucidity, surface properties and mechanical properties of the film. Due to the high purity of Sipernat 44, a discolouration cannot be observed, even the roll does not show this undesired effect.

Toxicology:
Sipernat 44 can be considered unobjectionable with regard to physiology.

Physico-Chemical Data (standard values):
Medium size of aggregates: 3-4 um
Sieve residue (Mocker, 45 um): <0.1%
Tamped density: approx. 450 g/l
Drying loss (2 hrs. at 105C): measured values not consistent
Ignition loss (1 hr. at 800C): approx. 20%
pH-value (in 5% aqueous dispersion): approx. 11.8
SiO_2: approx. 42%
Al_2O_3: approx. 36%
Na_2O: approx. 22%
Fe_2O_3: approx. 0.02%

Luzenac America, Inc.: Talc as an Antiblock Additive:

Talc is commonly added to film resins to prevent blocking
in the resulting film. Concentrations from 1,000 to 10,000 ppm
are commonly used. Luzenac America offers selected talc products
that provide the optimal compromise between film property
retention and antiblock performance. A surface modified product
is also offered where melt fracture or interactions with fluorocarbon processing aids needs to be minimized.

Talc offers a safer alternative to silica antiblocks.

Talc is a white, crystalline, platy mineral and is the softest
of all minerals. It must be mixed in the polymer melt to insure
good dispersion.

MISTRON 400C:
    Median Particle Size (microns): 3.9
    Topsize (microns): 25-30
    Loose Bulk Density (lbs/ft3): 9-13
    Surface Modified: No

JETFIL 575C:
    Median Particle Size (microns): 3.4
    Topsize (microns): 25-30
    Loose Bulk Density (lbs/ft3): 55-65
    Surface Modified: No-Compacted

STELLAR 510:
    Median Particle Size (microns): 3.8
    Topsize (microns): 25-30
    Loose Bulk Density (lbs/ft3): 9-13
    Surface Modified: No

Stellar 510F:
    Median Particle Size (microns): 3.8
    Topsize (microns): 25-30
    Loose Bulk Density (lbs/ft3): 9-13
    Surface Modified: Yes

Mistron PE:
    Median Particle Size (microns): 2.2
    Topsize (microns): 20-25
    Loose Bulk Density (lbs/ft3): 9-13
    Surface Modified: No

Witco Corp.: Additives for Polyolefins: Slip and Antiblock Agents:

In film, Kemamide E Ultra fatty amide is the industry's leading slip agent. It offers not only excellent slip characteristics, but also some antiblock effectiveness. Kemamide OR and Kemamide U fatty amides have faster bloom time and are also excellent slip additives, but have higher volatility than Kemamide E Ultra. Kemamide P-181 and E-180 fatty amides are extremely low-volatility slip agents for use with high processing temperatures.

Kemamide B, S, W-40 and W-20 fatty amides are organic antiblock agents. They do not contain silicas or other minerals, so they have less effect on film haze. In some resins, they can also provide slip effects.

In certain applications, only a moderate level of slip is desired. Kemamide E Ultra, Kemamide OR, and U fatty amides are such effective slip agents that slight variations in additive concentration can have a large effect on slip. Kemamide E570 fatty amide provides moderate slip without being adversely affected by small variations in additive concentration.

Kemamide E Ultra:
FDA Sanctioned: Yes
Antiblock/Slip: EVA Modified PE/UHMWPE
Antiblock/Mold Release/Slip: HDPE, LDPE, LLDPE, Polypropylene
Kemamide U, OR, O:
FDA Sanctioned: Yes
Slip: EVA Modified PE/UHMWPE
Rapid Slip/Slip: HDPE, LDPE, LLDPE, Polypropylene
Kemamide E-180:
FDA Sanctioned: Yes
Slip: EVA Modified PE/LLDPE/Polypropylene
Kemamide P-181:
FDA Sanctioned: Yes
Slip: EVA Modified PE/Ionomer/LLDPE/Polypropylene
Kemamide S:
FDA Sanctioned: Yes
Antiblock/Slip: EVA Modified PE/Polypropylene
Antiblock: LDPE/LLDPE
Kemamide B:
FDA Sanctioned: Yes
Antiblock: EVA Modified PE/LDPE/LLDPE/Polypropylene
Foaming Aid: Foamed PE
Kemamide W-20:
FDA Sanctioned: Yes
Antiblock: LLDPE/Polypropylene
Antiblock/Slip: EVA Modified PE/LDPE
Kemamide W-40:
FDA Sanctioned: Yes
Antiblock: EVA Modified PE/LDPE/LLDPE
Kemamide E570:
FDA Sanctioned: Yes
Medium Slip: LDPE/LLDPE/Polypropylene

Trade Name Index

Trade Name	Supplier
ACRAWAX	Lonza
ADMEX	Harwick
AFFLAIR	EM Industries
AMOCO	Amoco
ARC YELLOW	DayGlo
ARGUS	Witco
ARISTECH	Aristech Chemical
ARMOSLIP	Akzo Nobel
ATMOS	Witco
ATMUL	Witco
AURORA PINK	DayGlo
AXEL	Axel Plastics Research Labs
BENZOFLEX	Harwick
BISOFLEX	Inspec
BLACK PEARLS	Cabot
BLAZE ORANGE	DayGlo
BYK	Byk-Chemie
CHEMSTAT	Chemax
CHLOROFLO	Dover Chemical
CITROFLEX	Morflex
CLARION	Sun Chemical
COLCOLOR	Degussa
CORONA MAGENTA	DayGlo
CRODAMIDE	Croda Oleochemicals
DAYGLO	DayGlo
DAZZLE COLORS	DayGlo
DERUSSOL	Degussa
DIMUL	Witco
DOW	Dow Chemical
DOW CORNING	Dow Corning
DRAKEOL	Penreco
DRAPEX	CK Witco
DUOPRIME	Harwick
DURASTRENGTH	Atofina
DYNACOLOR	Engelhard
DYNAMAR	Dyneon
ECONOMIST	Slide Products
EEONOMER	Eeonyx
ELFTEX	Cabot
EPOXEASE	Slide Products
ESACURE	Sartomer
ESPEROX	Witco
EVERFLEX	Harwick
E/Z PURGE	Elm Grove Industries

Trade Name	Supplier
FILLEX	Intercorp
FIRE ORANGE	DayGlo
FLONAC	Eckart America
GLYCOLUBE	Lonza
GRAPHTOL	Clariant
HARSHAW	Engelhard
HATCOL	Hatco
HERCOFLEX	Harwick
HEUCO	Heucotech
HEUCOPHOS	Heucotech
HEUCOROX	Heucotech
HI-LITE	Engelhard
HI-MOD	Franklin Industrial
HORIZON BLUE	DayGlo
HYSTRENE	Witco
JAYFLEX	ExxonMobil
JENKINOL	Acme-Hardesty
JETFIL	Luzenac America
KEMAMIDE	Witco
KEMAMINE	Witco
KEMIRA	Kemira
KEMOLIT	Intercorp
KETJENBLACK	Akzo Nobel
K-FLEX	BF Goodrich Kalama
KRONOS	Kronos
LANKROFLEX	Akcros
LEXOLUBE	Inolex
LICA	Kenrich Petrochemicals
LUBRIOL	Morton Plastics
LUBRISTAB	Morton Plastics
LUBRISTAT	Morton Plastics
LUMOGEN	BASF
MAGNA PEARL	Engelhard
MASTERCOLOR	Eckart America
MASTERSAFE	Eckart America
MEARLIN	Engelhard
MERROL	Harwick
METABLEN	Mitsubishi Rayon
METEOR	Engelhard
METEOR PLUS	Engelhard
MICAFLEX	Pacer Technology
MISTRON	Luzenac America
MOGUL	Cabot
MOLDWIZ	Axel Plastics Research Labs
MONARCH	Cabot
MORFLEX	Morflex

Trade Name	Supplier
NEON RED	DayGlo
PALATINOL	BASF
PAROIL	Dover Chemical
PENRECO	Penreco
PHOSFLEX	Harwick
PLASTICHLOR	Harwick
PLASTOMOLL	BASF
POLYCIZER	Harwick
PRINTEX	Degussa
REGAL	Cabot
RIT-CIZER	Rit-Chem
ROCKET RED	DayGlo
SANTICIZER	Solutia
SARCAT	Sartomer
SATURN YELLOW	DayGlo
SIGNAL GREEN	DayGlo
SIPERNAT	Degussa
STANDART	Eckart America
STAN-LUBE	Harwick
STAN-PLAS	Harwick
STAPA	Eckart America
STELLAR	Luzenac America
SUDAN	BASF
TEGO	Tego Chemie Service
THERMOPLAST	BASF
TIONA	Millenium Inorganic
TIPAQUE	Ishihara
TRONOX	Kerr-McGee
UNIFLEX	Arizona Chemical
UNIPLEX	Unitex
VINYLUBE	Lonza
VISCOBYK	Byk-Chemie
VULCAN	Cabot
WATCHUNG	Clariant
XTEND	Axel Plastics Research Labs

Suppliers' Addresses

Acme-Hardesty Co.
1787 Sentry Parkway West
Blue Bell, PA 19422
(215)-591-3610/(800)-223-7054

Akcros Chemicals America
500 Jersey Ave.
P.O. Box 638
New Brunswick, NJ 08903
(732)-247-2202

Akzo Nobel Chemicals Inc.
300 South Riverside Plaza
Chicago, IL 60606
(312)-906-7500/(800)-828-7929

Amoco Chemicals
BP Amoco Chemicals
200 East Randolph Drive
Chicago, IL 60601
(630)-434-6200/(800)-621-4567

Aristech Chemical Corp.
Neville Island Plant
Pittsburgh, PA 15225
(412)-778-3300

Arizona Chemical Co.
P.O. Box 550850
Jacksonville, FL 32255
(904)-785-6700

Axel Plastics Research Labs
P.O. Box 770855
Woodside, NY 11377
(718)-672-8300/(800)-332-2935

Aztec Peroxides Inc.
555 Garden St.
Elyria, OH 44035
(440)-323-3112

Baerlocher USA
3676 Davis Road NW
P.O. Box 545
Dover, OH 44622
(330)-364-6000

BASF Corp.
36 Riverside Ave.
Rensselaer, NY 12144
(518)-472-8300/(877)-747-1857

Byk-Chemie USA
524 S. Cherry St.
Wallingford, CT 06492
(203)-265-2086

Cabot Corp.
157 Concord Rd.
P.O. Box 7001
Billerica, MA 01821
(617)-890-0200

Chemax Inc.
P.O. Box 6067
Greenville, SC 29606
(803)-277-7000/(800)-334-6234

Clariant Corp.
500 Washington St.
Coventry, RI 02816
(401)-823-2208

Croda Universal Inc
4014 Walnut Pond Drive
Houston, TX 77059
(281)-282-0022

DayGlo Color Corp.
4515 St. Clair Ave.
Cleveland, OH 44103
(216)-391-7070

Degussa Corp.
65 Challenger Rd.
Ridgefield Park, NJ 07660
(201)-641-6100

D.J. Enterprises Inc.
P.O. Box 31366
Cleveland, OH 44131
(216)-524-3879

Dover Chemical Corp.
3676 Davis Rd, NW
P.O. Box 40
Dover, OH 44622
(330)-343-7711/(800)-321-8805

Dow Chemical USA
Midland, MI 48674
(800)-447-4369

Dow Corning Corp.
Box 0994
Midland, MI 48686
(517)-496-6000

Dyneon
6744 33rd St. N
Oakdale, MN 55128
(612)-737-6700/(800)-863-9374

Eckart America
72 Corwin Dr
Painesville, OH 44077
(440)-354-0400/(800)-556-1111

Elf Atochem North America
Atofina Chemicals Inc
2000 Market St.
Philadelphia, PA 19103
(800)-446-2800/(610)-878-6658

Elm Grove Industries, Inc.
P.O. Box 659
Elm Grove, WI 53122
(414)-797-9244/(800)-797-9244

EM Industries, Inc.
7 Skyline Drive
Hawthorne, NY 10532
(800)-EM-PEARL

Engelhard Corp.
101 Wood Ave.
P.O. Box 770
Iselin, NJ 08830
(732)-205-5000

ExxonMobil Chemical Co.
P.O. Box 3272
Houston, TX 77253
(713)-870-6000

Ferro Corp.
7050 Krick Rd.
Walton Hills, OH 44146
(216)-641-8580

Franklin Industrial Minerals
P.O. Box 729
1469 S. Battleground Ave.
Kings Mountain, NC 28086
(704)-739-1321/(800)-290-2443

Genesee Polymers Corp.
Fenton Rd.
P.O. Box 7047
Flint, MI 48507
(810)-238-4966

B.F. Goodrich Kalama Inc.
1296 Third St. NW
Kalama, WA 98625
(360)-673-2550

Halstab
3100 Michigan St.
Hammond, IN 46323
(219)-844-3980

Harwick Standard Distribution
60 S. Seiberling St/Box 9360
Akron, OH 44305
(330)-798-9300

Hatco Corp.
King George Post Rd.
Fords, NJ 08863
(908)-738-3509/(908)-738-3646

Heucotech Ltd.
99 Newbold Rd.
Fairless Hills, PA 19030
(800)-HEUBACH

Inolex Chemical Co.
Jackson & Swanson Sts.
Philadelphia, PA 19148
(215)-271-0800/(800)-521-9891

Inspec UK, Ltd.
Charleston Rd.
Hardley, Hythe
Southampton, UK
SO45 3ZG

Intercorp Inc.
3628 W. Pierce St.
Milwaukee, WI 53215
(414)-383-2021/(800)-532-6303

Ishihara Corp.
600 Montgomery St.
San Francisco, CA 94111
(415)-421-8207

Kemira Pigments Inc.
P.O. Box 368
Savannah, GA 31402
(912)-652-1000

Kenrich Petrochemicals Inc.
140 East 22nd St.
P.O. Box 32
Bayonne, NJ 07002
(201)-823-9000/(800)-LICA-KPI

Kerr-McGee Chemical Corp.
Kerr-McGee Center
P.O. Box 25861
Oklahoma City, OK 73125
(800)-654-3911

Kronos Inc.
5 Cedar Brook Drive
Cranbury, NJ 08512
(609)-860-6230

Lonza
Algroup Lonza
17-17 Route 208
Fair Lawn, NJ 07410
(201)-794-2400

Luzenac America
9000 E. Nichols Ave.
Engelwood, CO 80112
(303)-643-0400/(800)-325-0299

Mitsubishi Rayon Co., Ltd.
520 Madison Ave.
New York, NY 10022
(212)-605-2406

Morflex Inc.
2110 High Point Rd.
Greensboro, NC 27403
(336)-292-1781

Morton Plastics Additives
2000 West St.
Cincinnati, OH 45215
(513)-733-2100

Neville Chemical Co.
2800 Neville Rd.
Pittsburgh, PA 15225
(412)-331-4200

Penreco
138 Petrolia St.
Kars City, PA 16041
(412)-756-0110/(800)-245-3952

PPG Industries
One PPG Place
Pittsburgh, PA 15272
(412)-434-3131/(800)-243-6774

Rit-Chem Co. Inc.
P.O. Box 435
Pleasantville, NY 10570
(914)-769-9110

Rohm and Haas Co.
100 Independence Mall West
Philadelphia, PA 19106
(800)-922-8765

Sartomer Co.
Oaklands Corporate Center
502 Thomas Jones Way
Exton, PA 19341
(610)-363-4100/(800)-SARTOMER

Slide Products Inc.
430 S. Wheeling Rd.
P.O. Box 156
Wheeling, IL 60090
(847)-541-7220/(800)-323-6433

Solutia Inc.
10300 Olive Blvd.
P.O. Box 66760
St. Louis, MO 63166
(314)-674-1000

Spectrum USA Inc (Millenium)
200 International Circle
Hunt Valley, MD 21030
(410)-229-4400

Stoner Inc.
1070 Robert Fulton Hwy
P.O. Box 65
Quarryville, PA 17566
(717)-786-7355/(800)-227-5538

Sun Chemical Corp.
5020 Spring Grove Ave.
Cincinnati, OH 45232
(513)-681-5950/(800)-543-2323

Tego Chemie Service GmbH
914 E. Randolph Rd.
Hopewell, VA 23860
(804)-541-8658/(800)-446-1809

Unitex Chemical Corp.
P.O. Box 16344
520 Broome Road
Greensboro, NC 27406
(336)-378-0965

Witco Corp.
Polymer Chemicals Group
One American Lane
Greenwich, CT 06831
(203)-861-6279/(800)-494-8737

Witco Corp.
Polymer Additives Group-Vinyl
8 Wright Way
Oakland, NJ 07436
(201)-337-2036